MW00908199

THE FINAL SALUTE

By

Kathleen M. Rodgers

Published by Leatherneck Publishing
A Division of Levin Publishing Group
http://www.leatherneckpublishing.com

Publisher: Lt Col H. Neil Levin, USMC (ret)

Copyright © 2008 by Kathleen M. Rodgers. All rights reserved

No part of this publication may be reproduced, stored in a retrieval system, or transmitted in any form or by any means, electronic, mechanical, photocopying, recording, scanning, or otherwise, except as permitted under Section 107 or 108 of the 1976 United States Copyright Act, without either the prior written permission of the publisher, or authorization through payment of the appropriate per-copy fee to the Copyright Clearance Center, Inc. 222 Rosewood Drive, Danvers, MA 01923, (978) 750-8400, fax (978) 750-4470.

Request to the Publisher for permission should be addressed to the Legal Department, Leatherneck Publishing, 3685 Vista Campana N. Unit 36, Oceanside, CA 92057 (760) 967-9575 fax (760) 967-9587 E-mail: per@leatherneckpublishing.com

The Leatherneck Publishing logo and related trade dress are trademarks or registered trademarks of Leatherneck Publishing, a Division of Levin Publishing Group in the United States and other countries, and may not be used without written permission.

Edition: 10 9 8 7 6 5 4 3 2 1 SAN: 256-8799

Library of Congress Control Number: 2008935845

ISBN-10: 0-9820892-0-1
ISBN-13: 978-0-9820892-0-0

Cover Design by Robert S. Jones
Author Photo: "Simply Black and White by Christina."
Printed and Bound in Mattoon, Il 61938
United States of America.
This Book is printed on Acid Free Paper

ACKNOWLEDGEMENTS

When I started this novel sixteen years ago, I had no idea how long the journey would take. Along the way, I have stumbled into roadblocks and detours, but somehow I kept going. Then one day I bumped into an angel who believed in my story. That angel is Neil Levin, Publisher of Leatherneck Publishing. I am grateful for his faith in me and the gift of his friendship.

A writer dreams of working with an editor who is firm but kind, fun but never flighty, and is always one step ahead of the writer. Nancy King at Leatherneck Publishing is that kind of editor. She took me by the hand and pulled me out of some rough spots.

I am forever indebted to my good friend Anita Robeson, a professional copyeditor who took me on as a charity case in the early days when this story was a work in progress.

Thanks also to Anita Peters, my counselor and teacher at Tarrant County College, Northeast Campus. Anita believed in me when I didn't believe in myself. To TCC American Literature Professor Cindy Baw, who told me I had talent when I thought that talent had dried up. Thanks to Suzanne Frank, head of the continuing education writing program at SMU. I am eternally grateful to the late Bill Southard, a novelist and newspaperman, who gave me my first professional break when I was nineteen. Bill said he didn't have an opening then hired me anyway.

To my soul sister, Rhonda Revels, thanks for your unending support. My ship finally came in.

My deepest love to my parents, Patricia Lamb Doran and Richard Leroy Doran. Thanks also to my sisters Laura and Jo-Lynda, fellow military wives, and also to my brothers Patrick and Richard, and our late brother, Larry.

Special love and thanks to my extended family and many friends, who always believed in me as a person, not just a writer.

Last but not least, to my husband Tom, and my two handsome sons, Thomas and J.P. It's been a long haul fellas. You all deserve medals.

DEDICATION

To Tom
whose ghosts first inspired this story.

And in memory of fallen friends,
too numerous to mention here.

TABLE OF CONTENTS

Mrs. Husson wrote this poem after her first husband, Captain Roy Westerfield, was killed in a plane crash February 6, 1980.

TAPS

By

Maryellen Husson

His best friend handed her
the carefully folded flag.
She held it to her heart
as he stepped back and saluted.

Soon would come
the final salute;
Taps would be sounded;
a chapter in her life would end.

She remembered the times
he had played this melody
for others at rest,
every note on his trumpet
strong and clear.

Bracing herself,
barely breathing,
she clutched the flag closer,
determined to face this moment
with dignity.

Then, like a rainbow,
the music wrapped around her
and the heaviness
in her heart lifted.

The music became his;
playing Taps one more time
just for her.

PROLOGUE

Williams Air Force Base, Arizona, 1971—On the first day of pilot school, a silver-haired colonel stood before a class of eager students and welcomed them with his standard opening speech.

"One fourth of you gentlemen will not be good enough to get through this program," he said. "As for the rest of you, the ones who don't wash out." His eyes scanned the room. "Take a look at the guy on your left, and the guy on your right. In twenty years of flying, one of you will be dead." He cleared his throat. "That's one out of three, gentlemen."

Tuck Westerfield, a second lieutenant from New Orleans, glanced at the cocky blond on his right. Jeff "Sweenedog" Sweeney winked back, then flashed Tuck a brassy grin.

PART I

Another Dead Pilot

Chapter One

Beauregard Air Force Base, Louisiana, 1990—"Sweenedog, you dumb son-of-a-bitch," Tuck slurred into the sticky darkness.

His shoulders sagged as he stumbled down Flight-line Road at midnight. His flight suit felt damp with sweat. He thought he smelled like a goat.

Each time Tuck strayed too far left or right, he forced himself back toward the center of the road. A fighter pilot, even a drunk fighter pilot, would never allow himself to fall into a ditch. Ditches were for pussies.

"Sweenedog, you dumb, stupid, son-of-a-bitch," Tuck snarled again, as if Sweenedog could actually hear him. "Why'd you have to go and get your ass killed..." Tuck's voice trailed off, swallowed by the night. *Jesus. What the hell happened, Sweenedog? That's the problem with ghosts,* Tuck thought. *They never answer.*

Under the glow of a street lamp, he plodded along, frowning occasionally at his clunky combat boots that resisted his efforts to walk a straight line. He wanted to get home, drop into bed next to his wife, and sleep the sleep of the dead, where even the ghosts took the night off.

When he left for work that June morning, he'd been a

sprightly forty-two. One lousy phone call aged him and now he felt a thousand years old.

A Friday afternoon and evening slamming down beers at the Officer's Club left him in no condition to drive. His eyes stung like hell, and the blue flight cap with the perfect crease down the middle—the one he usually wore with such precision—hung lopsided off his head.

"Sweenedog," he hiccupped. "Tonight, I don't exactly look like a poster boy for the Air Force." He stopped long enough to adjust his cap, then shuffled forward. He could see the street lights of base housing in the distance. "Gotta keep going," he urged his long, weary legs. "Not much farther."

Up the road, a pair of horses lingered along the side of the corral. The base stables sat back from Flight-line Road in a secluded nook bordered on one side by the dense banks of a bayou crawling with water moccasins and crawfish.

On the other side of the stables, a trio of enormous aircraft hangers dominated the view. Facing the hangers, an extensive taxiway led to two runways that seemed to stretch from yesterday to tomorrow. Several dozen A-10 Thunderbolt II aircraft rested in a roped-off area in front of the hangers, their jet engines cooled after a busy day of flying. That night, like every night, a car's headlights illuminated the area as security police made their way up and down the rows of multimillion-dollar jets. Finding nothing amiss, the car drove away.

Another set of headlights lit the side of the corral. A four-door sedan, dark blue with a white top, rolled by slowly and parked behind the barn. The headlights darkened before the driver rolled down the windows and turned off the engine.

Tuck felt the fog lift inside his head. *Hmm, maybe this long walk home has done me some good. I'll be sober by the time I get home. The last thing I want is to stumble through the front door and wake the kids or piss off Gina.* He veered off the road to take a whiz at the stables. Leaning against the fence, Tuck unzipped his flight suit down to his crotch and waited for the slow trickle, then the flood. As the sound of urine splashed on the ground, he thought he heard someone laugh.

"Sweenedog, old pal," he sighed. "I'm too old for this crap. Now I'm hearing voices. At least you won't have to worry about getting senile."

That morning at the squadron, as Tuck headed out the door to fly, an old friend from pilot school called with bad news from across the Atlantic. Jeff Sweeney "bought the farm" in England—his F-111 crashed into a farmer's field a couple of miles from the base. The U.S. Air Force investigated the incident and couldn't reach the widow for comment.

Same old story, Tuck thought, and he zipped up his flight suit. *Change the names, change the airplanes. In this business, death has a revolving door.*

Back on the road, he heard a woman giggle. A second later, he heard a man's low, coarse laugh.

Then, silence.

He listened to the sound of his own breathing and heard a horse nuzzle its moist nose against a fence, a frog croaked in the bayou, a cricket chirped far away in the woods.

Tired, Tuck rubbed his eyes and gathered himself up to head home. He could still see Sweenedog's shit-eating grin their first day of pilot school at Willie. *It was always the other guy who was going to die. Never you. You'd never be the one in three.* After that day, Tuck and Jeff Sweeney had become best friends. They flew F-4s together in Vietnam. Those days seemed like yesterday, but they were a lifetime ago. Tuck shuffled off, his leather boots crinkled under his weight. *In the morning, Sweenedog's memory will be gone*, he reminded himself.

To keep from going crazy, Tuck zipped up the memories, bagged them away and buried his grief. Nineteen years of flying fighters and fifty dead friends taught him that.

He hadn't gotten far when the laughter started again. This time, he could tell it came from behind the barn. Curious, he went to check it out.

Rounding the corner, he froze at the sight of the dark blue sedan with a white top parked behind the barn, and he knew the only person authorized to drive it, Colonel Glen P. Dennison, the base's highest ranking officer.

Tuck recalled seeing the wing commander at the club that evening in unusually high spirits, considering Mrs. Dennison was home recuperating from a recent hysterectomy.

Cautiously, he inched toward the car. When the car's dome light flickered on, Tuck sobered.

The back of Colonel Dennison's head pressed against the driver's door and he uttered a loud moan. Tuck crept forward. A few feet from the door, he stepped on a twig.

It broke with a snap and a woman's head popped up. Even in the dim light Tuck recognized the woman with short, blonde hair. Captain Linda Garrett, Colonel Dennison's wing executive officer.

Tuck's mind went blank.

A second later, a beam of light flooded the area, blinding him. By the time he could see again, an Air Force security policeman with gun drawn got out of a patrol car.

Another car pulled up, and two men got out, one, skinny, bald, and clad in dress blues, the other, a barrel-chested pilot with a shock of white hair.

The pilot with white hair thundered forward. "What the hell...?" He stopped short, gawking in disbelief.

Chapter Two

Friday afternoon, a week later, Tuck set his beer on the patio table.

"Damn dog," he growled and picked up the pellet pistol, stuffed a plastic grocery bag into a pocket of his rumpled flight suit and tiptoed into the yard.

After a long day at work, and having to dodge the wing commander all week, Tuck was in no mood to scoop up his neighbor's dog crap. That's why he and Gina didn't have a dog. They had boys.

"Don't move," he whispered. He crept toward the swing set and aimed the pistol at his next-door neighbor's fluffy Pekingese.

Fingering the trigger, Tuck moved in for the kill.

The dog hunkered down and looked the other way. The late afternoon sun bounced off the slide and into Tuck's eyes. He blinked, then squinted and anticipated his next move. Any second now, he expected to hear the swoosh of air when the pellet left the chamber and the high-pitched yip-yap of an indignant dog.

At the last second, he hesitated. His dead grandmother's chipper voice rattled around inside his head. *"Quit acting like some vigilante, Tucker Boy, and call the authorities."*

Tuck sighed. Since last week, he didn't trust Beauregard's authorities.

He steadied his hand on the gun, determined to fire. But another voice stepped in, his father's voice, the voice of reason: *"Can't this be handled in a neighborly fashion, son?"*

Tuck wiped the sweat off his brow with the heel of his hand. He hadn't talked to his father in over a week, not since he heard about Sweenedog. He felt a pang of guilt and made a mental note to call his folks over the weekend. He couldn't tell his mother about Sweenedog's crash. She would only start to worry again.

Tuck swallowed. He had a briny taste in his mouth. The air reeked with the stench of the paper mill down by the Red River—its putrid, fetid breath seemed stronger today than usual. He almost forgot how rotten the paper mill could smell in the summer, when mixed with the humidity and the scent of the malignant vegetation that threatened to overtake the state of Louisiana.

Tuck glared at the dog. A month after the Sandfords moved in two years ago, Tuck went next door after work and asked them, kindly, to control their dog. When Major Big Sandy Sandford answered the door with a plug of chewing tobacco lodged in one cheek, he took one hard, contemptuous look at Tuck's flight suit and higher rank, nodded "uh-huh" a couple of times, then shut the door in Tuck's face.

From that day on, the dog made another deposit for Tuck to clean up. Obviously, Beauregard's newest Chief of Civil Engineering didn't like pilots much. Especially pilots who complained about his wife's dog.

Instead of turning the major and his wife into the housing office, which Tuck had every right to do, he devised his own scheme to get even. Today, the pellet pistol gave an added bonus.

The Pekingese got up then resettled itself.

Tuck started to squeeze the trigger then stopped. *You're just pissed off,* he thought. *Don't take your frustrations out on a helpless animal.*

Tuck closed his eyes and in a moment of weakness, Sweenedog's image swam into view. Tuck could see the big, gregarious Hollywood grin, the baby-blue eyes that twinkled when he brushed back a lock of blond hair and turned in his desk to wink at Tuck.

Now Sweenedog was dead, and the three towheaded Sweeney children didn't have a father. In the blink of an eye, Tuck banished Sweenedog back to that place in his brain where he didn't have to deal with things. He drew in a lungful of air and readied the trigger.

Suddenly, a crack-crack-pop noise exploded behind him. He jerked his head around and saw two teen-age boys race away from a smoldering fire ant mound.

"Hey," Tuck yelled at the retreating boys. Firecrackers continued to pop over the ant mound. By the time Tuck turned back toward his prey, the Pekingese scampered away, unscathed.

"Damn mutt," Tuck growled and headed to the swing set to clean up the mess—and to get some small measure of revenge.

He heard the dog's mistress calling from her back door.

"Baby. Where did you go, little man? Mama's got your din-din."

The dog sniffed up the base of a pecan tree nearby, where a squirrel darted to safety.

Tuck glanced up when Wynonna Sandford, a pudgy, peroxide blonde, clip-clopped across her yard in three-inch heels and skintight jeans with ankle zippers. Her big sagging breasts jiggled beneath her frilly blouse, knotted at the midriff.

"Don't shoot," Wynonna cried when she saw the gun in Tuck's hand. She stopped at the edge of her yard and threw up her arms, crisscrossing them in front of her like a shield.

Tuck grinned awkwardly at his neighbor. "Hello, Wynonna."

He worked up the courage to look her in the eye then glanced at her open-toed mules. Her stubby toes, painted purple, poked out like Vienna sausages.

Relaxing her arms, Wynonna propped a manicured hand on one ample hip. "Good gravy," she crowed then batted her bushy lashes at Tuck. "You nearly scared me to death."

Wynonna had one of those pretty faces that took a wrong turn in life and looked lost and forlorn half the time—behind all that make-up she slathered on. She drove a customized, purple Suburban and sold Purple Passion beauty products door to door. Because of her gaudy outfits, people said Major Sandford's wife looked more like a hooker than an officer's wife. She stood a few feet from the swing set, careful not to cross the invisible line between their yards. With no fences to mark the boundaries between houses on Beauregard, everyone, including Wynonna seemed to understand his or her territory.

Everyone but the neighborhood children and Wynonna Sandford's dog.

"What in the world are you doing with that gun?" she asked.

Tuck motioned toward her dog. "Warning that pug-nosed trespasser over there to stay out of my yard."

Wynonna's jaw dropped. "Baby wouldn't hurt a flea." She thrust both hands on her hips, and her chest heaved up and down, as if the air was too thick to breath. "My granny always said never trust a man with a dimple in his chin."

Instinctively, Tuck reached for the notch in his chin that caused him so much consternation as a boy. His own granny felt differently. He recalled the day he ran to her and complained he had a hole in his chin.

"Nonsense, Honey Boy. That's where God left his thumbprint." She shoved a pie in the oven then turned to wrap him up in her arms against her big floured apron.

Tuck's throat ached at the memory. Like he'd done with Sweenedog, he banished the old lady from his thoughts.

He gazed at Wynonna tottering in her vampish heels and flashy outfit.

He felt sorry for her, then changed his mind. How many times had the boys come running to him with dog poop on their sneakers and clothing? "It stinks, Daddy," Austin gagged.

"But why would you want to hurt Baby?" Wynonna asked, as if the words themselves drew pain.

Tuck shifted the gun to his other hand and took the plastic bag from his pocket. "I wasn't meaning to hurt your dog, just, well"—he shook out the bag and looked at her— "if you had kids, you'd understand."

Wynnona shuddered and hugged herself.

Tuck slipped the bag over his hand, then stooped over and scooped up the pungent mess. When he stood up, Wynonna had her nose in the air and a pinched expression on her face. She held out her thumb and forefinger like a pair of tweezers and waited for him to pass her the bag.

Instead of handing it to her, he cocked his arm like a center fielder ready to lob the ball home.

"What are you doing?" she gasped.

"Better call in the troops," he laughed. "You've got incoming." Tuck shot her a wink and launched the smelly brown missile high into the air, without letting go of the bag. The droppings sailed through the air and landed smack dab in the middle of Wynonna's lawn.

He squared his shoulders and tried to offer her the bag. "Dog shit," he declared with a triumphant grin, "the perfect weapon. It beats chemical warfare and it's biodegradable."

Wynonna's eyes darted wildly from the dark clumps in her yard to the pellet pistol in Tuck's hand.

"You old fool," she spat. "You aimed to kill Baby, or cripple him for life...all because of a little accident?"

"Accident my ass," Tuck laughed. "You've disregarded

Beauregard's leash law for two years. I wasn't going to kill your dog. I only wanted to scare the little scoundrel. Teach him some doggy etiquette."

Wynonna's lips formed an angry knot. "You're out of line." She thrust her ample bosom forward. "Big Sandy's right about you pilots. You've got no class."

"Maybe not," he said. "But if I had a dog, he'd be on a leash."

"You're an asshole," she blasted before stalking off, her spiked heels digging into the yard like nails.

"At least I'm a law-abiding one," Tuck called.

"Fat chance of that," she snorted over her shoulder. "You're an attempted dog-murderer, Colonel Westerfield."

Tuck chuckled and watched her go. But when Wynonna scooped up the Pekingese and cuddled him to her bosom, Tuck felt a stab of guilt as she stormed into her house and slammed the door.

"Honey…?"

Tuck whirled at the sound of his wife's voice.

Gina, a tall brunette in a pair of cut-offs and a T-shirt, sounded more curious than alarmed. She stepped onto the patio in her bare feet, then turned to pull the sliding glass door behind her and started toward him. Her long, brown legs glistened in the late afternoon sun.

"What was that popping noise I heard?"

"Firecrackers," Tuck yelled.

At thirty-six, Gina moved with the ease of a woman comfortable with her beauty. She had high, prominent cheek bones and a figure fit from running. The only evidence she'd given birth were the two brown-haired boys who tagged along beside her everywhere she went. Her mouth was pretty and vulnerable—until she got mad.

"I wouldn't come out here barefoot if I were you," he cautioned and gestured to several spots in the yard. "It's like walking into a minefield."

Gina gingerly picked her way through the yard, suspect of every blade of grass, every small, curly twig. When she spotted the gun, she shrieked.

Tuck held up the pellet pistol. "I found it out in the shed, in one of those boxes we never unpacked from the last move."

Gina shrank back. "I thought you got rid of that thing." She glanced back at the house. "I don't want the boys to see it."

Tuck shook his head at her logic. He flew jets capable of mass destruction, yet Gina worried about a pellet pistol?

She eyed him suspiciously. "What are you doing with it out here in the middle of the yard? I don't see any targets set up."

Tuck dangled the plastic bag in front of her and motioned next door with the pistol. "The next time the Sandford's pooch traipses over here, he's history."

Gina's eyebrows shot up. She looked around, half expecting someone to show up and arrest Tuck. "For Pete's sake, Honey, you're the Chief of Safety."

Tuck grinned sheepishly, then stared at his feet.

When he wasn't flying, he sat behind a desk at wing headquarters, pushed papers, or went out on inspections. Being Chief of Safety wasn't a bad job, only another square to fill. Another rung up the ladder in his quest to command a squadron.

Gina's voice skittered on the edge of panic. "If Colonel Dennison gets wind of this, he'll take your name off the squadron commander's list."

Tuck set his jaw. "Now don't go jumping to conclusions. I haven't done anything wrong." He gestured toward the large ant mound on the side of the house.

Gina looked over and wrung her hands and waited for him to explain himself.

"Some kids got their jollies by lighting firecrackers to blow up fire ants." He paused. "Trust me, Babe, I never got a shot. The dog got spooked and ran off."

"Those stupid kids." Gina's forehead knotted in a scowl. "All they did was get the ants stirred up." She glanced next door. "You better hope to God Wynonna doesn't find out about your gun. Big Sandy'll kick your ass."

"Too late," he chuckled. He waggled his brows at the pellet pistol. "Boy is she dumber than a bag of rocks. She actually thought I would kill her dog with this thing."

Gina's mouth flew open. "Don't tell me you were really going to shoot it?"

Tuck shrugged. "Who knows? The point is, I didn't."

"Then why didn't you set her straight?" Gina jammed her hands into her pockets. Her long slender arms, like her legs, remained stiff at the moment.

Because I'm dying inside, Tuck wanted to tell her. *And I'm tired. And I'm a fool. And I don't know who to trust anymore.* Instead he shrugged, and admired his wife of fifteen years and waited for her to dish out her reprimands.

Today, Gina's chestnut mane was swept up in an old-fashioned bun, but when feeling sexy or sassy, she wore it down. Tuck liked to think of Gina's hairstyle as a weather vane.

"But the dog's an innocent creature." A storm brewed in her hazel eyes.

"Hey, life's a bitch and then you die," Tuck grumbled.

Gina stuffed her hands deeper into her pockets. "That's the same thing you said about Sweenedog's accident." She took a deep breath. "What's gotten into you?"

"Look, I've got a lot on my mind, okay?" He tried to brush her off.

"You keep to yourself too much." She squished tufts of grass between her toes. "It's not healthy."

"Gina, I'm fine. Now get off my back."

He wanted her to turn away. Instead, she withdrew her hands from her pockets and grasped his shoulders.

"This isn't about Wynonna's dog is it?" she whispered

and forced him to look at her. "It's about Jeff Sweeney."

Tuck couldn't escape her stare. Her eyes tapped into his soul and siphoned his thoughts. *Another good friend and fighter pilot was dead either because he fucked up or his airplane malfunctioned.* With no where to run, up went the stoic mask. He clammed up and stared into her eyes.

To add to his pain, Tuck was told to keep a tight lip on what he had seen at the stables the other night or suffer the consequences. The first harassing phone call came shortly after the white top deposited him safely at home after one a.m. He caught the phone on the first ring without arousing Gina.

"If you expect to take command of a squadron any time soon, you better play by Dennison's rules and keep your mouth shut. Don't fuck with him, Tuck—he'll stomp your dick in the dirt. If you screw up and spill your guts, we're all ruined. At least Dennison wasn't diddling some boy like that colonel at Ramstein. By the way, I left your car keys under the clump of marigolds by the front door."

Sure enough, a peek out the window informed Tuck his white '72 Corvette, which he left in the O'Club parking lot, had materialized in front of his quarters.

"Tuck." The sound of Gina's voice snapped him out of his fog.

He blinked.

"You're looking straight at me but you're a million miles away," she frowned. "For God's sake, Tuck, you're allowed to grieve." She studied him for a moment. "Is that what they taught you in pilot school? To box up your emotions and never deal with them? Is that another prerequisite to becoming a fighter pilot, along with topnotch eyesight and quick reflexes?"

Tuck laughed to cover the pain. "You forgot about the big watch and little dick."

Gina rolled her eyes. "And always trying to get laid for free," she stole his thunder. A favorite slogan among fighter pilots.

"Come on, Tuck, be serious. You cram too much stuff in that head of yours."

"Hey, it comes with the territory." He shrugged.

"You're impossible," she sighed and started to back away. "Whatever is going on with you, don't try to carry it alone. We're in this together." She gave him a hopeful smile.

He watched her walk back toward the house, admiring the wiggle in her firm bottom. At the patio table, she turned. "Don't forget, you promised to make the boys ice cream."

Tuck glanced at the shed, where he kept the ice-cream maker then noticed his two young sons peering at him through the sliding glass door. Jesse, his three-year-old, clapped his hands and jumped up and down, excited about something. But six-year-old Austin cowered behind a curtain.

Tuck hung his head. He felt foolish standing there in the yard with the pistol in one hand and the bag in the other.

When he heard the back door slide open, he looked up. Jesse tumbled out and galloped toward him, fingers like an imaginary gun.

"Bang bang, Daddy," giggled Jesse. His chubby finger pointed at Tuck. "Bang bang, Daddy. You're dead."

Tuck froze, wishing he could disappear.

Gina poked her head outside and hollered at Jesse. "Get back inside this house, Buster, till Daddy puts down more poison. The fire ants'll eat you alive."

With an impish grin, Jesse turned and scrambled back inside.

Tuck started for the house and saw Austin duck behind the curtain. At the back door, Tuck tossed the bag into a trash barrel, laid the pistol on the patio table and stepped inside, feeling like the biggest heel in the world.

"Come here." His voice sounded scratchy when he went to kneel down. He tried to coax Austin from his hiding spot. "Daddy needs a hug."

At last, Austin stepped from behind the curtain and

slipped into his father's embrace. Tuck buried his face in Austin's cap of brown hair, breathing in all of his little boy smells of salt and sand and sun and sweat, as if somehow the very act itself could recapture his own youth, restore his faith in humanity and in himself.

This was Tuck's way: a warm hug, an unspoken reassurance, an apology even, for being a lousy role model at times.

After a moment, Tuck released his grip and got up. "You wanna help me make ice cream?"

Austin appeared to hesitate. "Dad?"

Tuck looked at him. "What is it, son?"

Austin hung his head. "Why were you going to shoot Mrs. Sandford's dog? Did Baby bite you?"

Tuck sighed and put his hands on Austin's shoulders. "No. It's that. Well, I'm tired of him using our yard as a bathroom."

Austin wrinkled his nose, peppered with freckles like his mother's. "He pooped in front of the slide again, didn't he?"

Tuck nodded. "I only meant to warn him so he wouldn't come back."

Austin pursed his lips and needed more time to think. He turned and glanced outside at the patio table. "Mom says that's a pellet pistol."

"That's right."

"So. Is that like a BB gun?"

"Yeah, sort of I guess."

Austin's eyes lit up. "Could you teach me how to shoot it?"

Gina walked into the room and gave Tuck the evil eye.

"We'll see." Tuck put an end to the discussion and moved for the door. To appease Gina, he picked up the pistol, took it around to the side of the house, and stashed it on a top shelf in a tiny storage closet, where the boys couldn't reach it. He locked the door and came back.

A few minutes later, Tuck and Austin hefted a battered

ice-cream maker, a left over from Gina's childhood, out of the shed and lugged it to the patio. Tuck didn't relish the idea of cranking the rusty handle.

Austin stood up, brushing his hands. "How come you don't buy one of those electric kind like they sell at the BX? That way your arm wouldn't get tired."

Tuck paused and looked at his son before he went to retrieve the rock salt.

"Good question. But that'll be your mama's decision."

"How come?"

Should Tuck tell the boy why his mother refused to part with an ice-cream maker with a wooden tub that was damn near rotten? Yet every year Gina crossed her fingers: "Try 'n keep it working through the summer, Honey." Each summer, Tuck prayed for the old thing to fall apart.

He decided to divulge half the truth. "It reminds her of Grandpa Loyd."

Austin crossed his arms. "Mom says Grandpa Loyd wants to sell you more life insurance."

Tuck threw his head back and laughed.

"Dad," Austin cocked his head to the side, "does that mean if you die Mom gets a lot of money from the insurance company?"

"Yeah. It's sort of like winning the lottery." Tuck ruffled the boy's hair.

Austin reached down and gave the handle a spin. "What's so special about this old piece of junk anyway?"

Tuck scratched his head. "When your mama was a little girl, your grandpa used to sit hunched over a stool and crank that very handle. Sometimes he'd let her sit on his lap and help."

Tuck left out the rest of the story. How Gina's father skipped out on his devoted wife and only child years later, and left them and his prized ice-cream maker for, in Gina's words: "a skinny, chain-smokin' broad with a lactose intolerance."

They went inside to stir the ingredients and get a bag of

ice from the deep freeze. Austin stopped halfway through the door and yanked at Tuck's sleeve.

What now, Tuck thought, looking down.

Austin craned his neck, like he looked up at a giant. "Are you sad because Colonel Sweeney crashed his airplane?"

"Yes," Tuck croaked, gazing into his son's green eyes.

Some things were easier to admit to a child.

The phone rang.

Chapter Three

"Who was that?" Gina glanced up from the refrigerator. She got the eggs for the ice cream.

Tuck held the phone in midair. He stood across from her in their outdated kitchen with its dark, worn cabinetry and stark-white appliances. A pot of potpourri simmered on the stove and the room smelled faintly of vanilla and cinnamon.

"Wrong number." Tuck held his breath and turned to hang up the phone.

"Another wrong number?" Gina raised her voice.

Tuck's heart thumped against his chest like a kettledrum and pounded out the rhythm of his father's deep, throaty words that rippled back to him across the canyon of time: *Don't ever lie to me, Tucker Boy. If a man lies to you once, you can never trust him again.* Tuck cringed.

Gina slammed the refrigerator door with her foot and brushed past him. She set the eggs next to the open *Southern Living* cookbook on the gold-speckled countertop.

Tuck helped himself to the boys' cherry Kool-Aid. Dropping down on a barstool, he drained half the liquid in one swallow, then set the glass down to rub his neck. Every time

Tuck hung up from one of these calls—where an arrogant voice threatened him if he blew Colonel Dennison's cover—a knot of tension flared up at the base of his skull. The caller's words still rang in his ears. *"Remember your career, Tuck. You've got a promising future ahead of you. Don't upset the apple cart by venting your spleen—say, during one of your weaker moments when you're home with that pretty wife of yours. Be a shame to drag her and the boys into this."*

Gina leaned against the counter with her arms crossed and watched him. "Tuck, that's about the tenth wrong number we've had this week."

Tuck drank his Kool-Aid. "We could get our number changed."

Gina squinted at him like he was a hot, sweaty, tight-lipped impostor in a Nomex flight suit. "Don't you think that's a little extreme?"

Tuck shrugged and emptied his glass. He felt guilty, looked away and rubbed his neck.

Gina mumbled something under her breath. Her knees creaked when she crouched down to lift a large ceramic mixing bowl from a bottom cabinet. She set the heavy bowl on the counter and whisked past him to the pantry for the evaporated milk and vanilla. When she returned, she plunked down on a barstool across from him and drummed her fingers on the counter. "What's wrong, you got a crick in your neck?"

Tuck plunged his fingers deep into the knot at the back of his neck while another lie snuck past his lips. "I guess I pulled it flying this morning."

Gina stopped drumming her fingers. "You want me to get you the heating pad?"

He shook his head.

"Who did you fly with?"

Tuck felt grateful for the change of subject. "Wheaties. We went to the range and back. Strafed a few targets... dropped some bombs."

Gina perked up. "So, how is my favorite red-haired lieutenant?"

"He flies a good airplane."

Roy "Wheaties" Wheaton was Tuck's protégé and the youngest lieutenant in the 428th; Tuck and Gina practically adopted him, along with his young wife, Sylvia.

Gina leaned forward, grinning mischievously. "Is he still in hot water for ripping his clothes off at Killer's bachelor party?" Killer was a captain in the 428th.

Tuck shrugged, surprised by Gina's sudden change in mood—and the way she could look like a full-grown woman and a little girl at the same time. Her eyes lit up and sparkled with some inner fire Tuck found sexy.

"All I know is young Wheaties learned you don't go skinny-dipping in the Officer's Club Pool when there's a full-colonel's wife around," Tuck said.

Gina leaned closer and laughed. "What other fighter pilot is crazy enough to sing *Onward Christian Soldiers* on top of the high dive, rip off his flight suit, and then execute a perfect swan dive?"

"Lots of them," Tuck replied. "Wheaties just got caught by an old battle-ax who wears her husband's rank. What about you?" he teased and twitched his brows. "Have you ever pulled rank?"

Gina wriggled her nose at him. "You're a light colonel, Tuck. You don't count."

At moments like these, Tuck could almost forget the complex mess of his life.

She pulled a bobby pin out of her hair and stuck it between her teeth. "Seriously, is Wheaties in a lot of trouble?"

"Let me put it this way. Colonel Dennison warned Wheaties that if he ever pulls another stunt like that one, he might as well pack his red-headed-ass back to Kansas and fly crop dusters with his loony uncle."

Gina plucked the bobby pin out of her mouth and wedged

it back in her hair. "Promise me one thing, Tuck. If you ever make full-bird, you won't act like a jerk?"

She hopped off the stool to go check on the kids.

Gina doesn't know the half of it, Tuck thought and stood up to stretch.

Colonel Dennison accused Wheaties of bad judgment—behavior unbecoming an officer. He heard his own words in his head. "*The kid just had too much to drink.*"

"*No excuse*," came the reply from the top.

Tuck and Wheaties were both given a verbal reprimand and threatened with an Article 15, non-judicial punishment. Obviously bad judgment didn't apply to those in command—even if they got caught.

Finally at five-thirty Tuck started the ice cream.

"Looks like we're out of chocolate syrup." Gina stood at the refrigerator door, sucking on a lemon drop. "Jesse's going to raise holy hell."

Tuck cracked open an egg when the phone rang. He stared at the yolk as it plopped into the bowl and jiggled in a lumpy pool of milk and sugar.

He could feel Gina's eyes on his back. His fingers sticky, he stood over the mixing bowl and waited for her to pick the phone up.

"Why don't you answer it?" Gina asked on the third ring.

Tuck didn't budge. He stood there in his flight suit and boots, his heart pounding. *A man lies to you once, you can never trust him again.*

"Who is it, Tuck?" Gina came toward him. "What do they want?"

Tuck wiped his hands on the towel and turned to face her. She was so close he could smell the lemon drop on her breath.

"What's going on, Tuck? Several times this week, whenever I pick up the phone, someone hangs up."

Tuck busied himself counting the tiny freckles sprinkled

across Gina's nose and upper cheeks. He was lucky to have her. Gina could have fled the small pond of Llano, New Mexico, for bigger and better things. Instead, she married him and became a full-time military wife.

The phone continued to ring.

"Are you in some kind of trouble?" Her eyes narrowed.

"Trouble?" He threw the question back at her.

"Damn it, Tuck. Answer the phone." She crunched down on the lemon drop.

Just then Austin bolted into the kitchen and plucked up the phone. He gave his parents a funny look.

After a short pause, he passed the phone to Tuck. "It's for you, Dad."

Tuck cleared his throat and lifted the phone to his ear.

***

Gina noticed the way the muscles flexed along the ridge of Tuck's jaw. His ruggedly handsome face, with the friendly squint lines around his green eyes—from too much flying into the sun—had twisted into an intense scowl.

She glanced at Austin. "Who is it?" she mouthed.

"Some lady," he shrugged and ran out of the room.

At last, Tuck cupped his hand over the mouthpiece. "It's Sally."

Gina frowned at the mention of Tuck's ex-wife.

"She's going to South America—missionary work." He removed his hand from the phone.

Gina listened to Tuck's voice, a low rumble of "uh huhs" as he stepped back in cadence, talking into the receiver to Sally like he'd never been away.

"At least it wasn't another wrong number," Gina said.

Tuck put his finger to his lips for quiet.

Furious, Gina turned and marched out of the kitchen, down the hall past Austin's room where the boys played with

Legos, and into her room, then slammed the door behind her.

Alone in her room, Gina choked back her suspicion. A bobby pin fell to the floor. Then another. Her hairdo drooped hopelessly to one side. She yanked out the rest of the pins and shook her hair out.

In the span of one week, a sick feeling crept back into Gina's life every time the phone rang. She trembled and remembered the day in ninth grade when she found a pair of lacy panties in her father's car. And the look on her dear mother's face after Gina ran into the kitchen, panties in hand, and blurted, "Daddy's screwing around on you."

Long before I found the panties, I heard the string of wrong numbers, the odd phone calls Dad couldn't explain. Or wouldn't, Gina thought, standing at the foot of her bed, glaring at the phone. *Maybe that isn't Sally on the phone. Maybe Tuck is lying and having an affair. Could Jeff Sweeney's death have sent Tuck into a tailspin of middle-age-crazy? Didn't Dad skip out on us shortly after his brother was killed in a car wreck?* A hundred different reasons crossed Gina's mind why a man Tuck's age might mess around. For her own peace of mind, Gina plucked a washcloth from a stack of clean clothes on the bed and padded over to the night stand to listen in on the extension.

"Jesus H. Christ, Sally. You're letting the kid go to Houston with a carload of teeny-boppers and no adult? She's only sixteen. Anything could happen on that stretch of road between Galveston and Houston."

"You haven't changed one bit, have you, Tucker," Sally said. "I see you're still taking the Lord's name in vain."

Gina bit her lip at the sound of Sally's voice. Relief flooded her, but an inner voice, a woman's intuition, told her to stay on the line. Tuck sounded more agitated than usual.

"It's a *Christian* rock concert," Sally said. "Instead of drugs and sex, they sing about Jesus. Besides, she'll have her guardian angel with her."

"Oh horseshit," Tuck growled. "Whatever happened to

plain ol' chaperones? You know, Sally, the flesh and blood kind. The kind that can step in and stop trouble before it starts."

Gina let the phone slip off her shoulder. The kid. Tuck and Sally's marriage might have been short in terms of time, lasting only two months in early 1974, but the marriage produced a child, Michelle—Sally and Tuck's honeymoon baby. Tuck was unaware of the pregnancy when Sally filed for divorce a few weeks later. Her grounds? She couldn't live with a heathen like Tuck, or be married to a man who willingly dropped bombs on Asian babies for a living.

Gina found Michelle a difficult child to love, although they hadn't seen her in years. Until Austin and Jesse came along, Gina bent over backwards trying to please Michelle the few times the girl came to visit.

Michelle, in turn, played tricks on her stepmother and said hateful things to Gina when Tuck wasn't around. Saddest of all, the girl regarded Tuck with quiet contempt.

Yet Gina wrote out the monthly child-support checks and mailed them to the Galveston address. Tuck rarely talked about Michelle—like she didn't exist—until Sally called out of the blue to remind him.

Gina held the phone to her ear, barely breathing.

"We're pretty cramped here," Tuck said. "But the kid can take over one of the boy's rooms. We'll work it out."

Gina slammed down the phone and stormed into the kitchen with her hands on her hips. "Have you lost your mind?"

Tuck covered the phone. "Calm down, Babe. I'll explain in a minute." He apologized to Sally and hung up. "Keep your voice down." He motioned toward the back of the house. "You'll scare the boys."

Gina walked over to the mixing bowl. "Michelle's last visit was a disaster." She picked up a wire whisk and broke the egg yolks.

Tuck came up behind her and touched her on the arms.

"You didn't have to sneak back there and eavesdrop."

Gina tensed. "You've never given me a reason until now."

Tuck took a deep breath and let go of her. "Michelle needs a place to stay for a while. Jesse can bunk with Austin."

"And you expect me to go along with this?" In a frenzy, Gina whipped the ice-cream mixture.

"I'm her father."

Gina whirled on him. "She hates me." She held the whisk and let the creamy white goo drip onto the floor. "Sometimes I think she hates you, too."

Tuck's jaw tensed. "She has nowhere else to go."

Gina sighed and cupped her hand under the whisk to catch the drips. A white glop splattered on the toe of Tuck's boot.

"Here, let me do that." Tuck took the whisk out of her hand.

"When's she coming?" Gina sighed.

"Next week I th—" Tuck broke off when the phone rang.

Gina glared at the phone, then at Tuck, and walked out of the room. She hiked down the hall to grab the boys and pull on her sneakers.

Austin put the finishing touches on a small Lego airplane when she came into the room. Jesse dumped a bucketful of Legos on the floor. Gina gulped at the pained expression in their eyes—at that look of fear and confusion when a child gets caught in the middle of bickering parents. Gina sucked in her breath as her own childhood flashed by. History repeated itself.

"Come on, fellas. Let's go for a walk." Her voice cracked as she tried to sound cheerful.

She scooped Jesse up in her arms and moved down the hallway.

"Where we going, Mommy?"

The phone trilled on and on.

"The playground."

Austin tagged along behind her and clutched his airplane. "Is Dad coming with us?"

"No. Daddy's busy."

"Too busy to answer the phone?" Austin asked in a puzzled voice.

***

Tuck poured the ice-cream mixture into the silver canister and inserted the dasher. He was on his way out back when he heard the front door bang shut.

Gina has a right to be angry, he thought. He put the metal canister into the tub and filled it with ice and salt. *A daughter I barely know will walk back into my life.*

He pulled her down from one of those pigeonholes in his brain and thought about the last time he saw her. She was a gangly, disrespectful pubescent with fly-away blonde hair and tiny buds sprouting on her flat chest. "Brace yourself, Tucker, she's changed," Sally warned.

Tuck put the lid on the freezer and went after Gina. "Babe," he called from the top of the sloping driveway, "leave the boys with me and go for a run."

"I ran this morning." Jesse remained on her hip. Austin scampered along beside her. He kept glancing back like he'd lost something.

Halfway down the driveway Tuck said, "Gina, please. I've got the ice cream started."

"Asshole," she hissed over her shoulder as they fled down the sidewalk. "Why didn't you answer the phone?"

"Affhole," Jesse copied her, waving a chubby hand at his father. "Affhole, affhole, affhole," he jabbered down the tree-line sidewalk of Bayou Way till they disappeared around the bend.

A curtain moved at the stately, red-brick house across the street. Tuck knew he was being watched. He glared back. *Don't fuck with me*, he thought.

The blue sedan with a white top was parked under the portico in the middle of the circular flagstone drive. A cobblestone path lead to a pair of imposing wooden doors, through which generals, governors, and foreign dignitaries passed from time to time.

Tuck turned to go in. As he headed up the driveway, he spotted a pile of Legos, a tiny fuselage, two wings, and a tail. Tuck rescued the Lego pilot from the flower bed, looked around for the tiny helmet, then abandoned his search and went inside. After reassembling the pieces, he stuck the plane on a shelf in the living room, next to an old clock.

Tuck swallowed the lump in his throat and turned away. Too bad Sweenedog and his airplane couldn't be pieced back together like that.

The phone jingled, but Tuck ignored it. A minute later it rang again.

This time, without hesitation, he strode angrily into the kitchen. A man could only take so much.

***

The playground buzzed with the chatter of children while Gina sat hunched over a wooden bench, chin in hand, kicking at a bed of pea gravel. The boys played nearby.

The rocks crunched in front of her and she jumped, startled, when a long shadow fell over her and blocked the six o'clock sun.

Jo-Ellen Hawkins, a high-school English teacher with cropped, ash-blonde hair, gazed down as her cherry-red lips peeled back in a generous grin. "I thought that was you," Jo-Ellen said in her native Oklahoma twang.

"Hi, Jo. I didn't see you walk up."

Clad in a long denim skirt and a plain white blouse, Jo-Ellen peered at Gina over the rim of her sunglasses. "What'cha broodin' about?" The strap of her purse cut into one shoulder as she plopped down on the bench beside Gina.

Gina turned and looked at her friend. "Did Tuck send you down here to check up on me?"

Jo-Ellen smoothed her skirt. "No, was he supposed to?"

Gina shook her head. "I just thought maybe he called you to come look for me."

Jo-Ellen took off her sunglasses and cleaned them with the hem of her skirt. "I was on my way home when I saw you sittin' over here all by your lonesome." She slipped on her glasses.

Gina met Jo-Ellen back in March in the gardening aisle of the BX. Jo-Ellen was married to Chief Master Sergeant Buzz Hawkins, a senior NCO in charge of all enlisted maintenance personnel on base. In an unspoken rule, the Air Force discouraged their friendship, because Gina's husband was an officer and Jo-Ellen's enlisted. Neither woman cared what the Air Force thought.

Gina pulled her hair to the side and gazed at the playground.

Built of wood and metal the year before by a team of base volunteers, the playground towered like a dwarf-sized fortress in an open field on the southern tip of the base, between officer and enlisted housing. A ten-foot-tall chain-link fence protected the children from the bayou and separated the military from the rest of society. In the distance, past flat fields of corn and cotton, rested the outskirts of Bolton, Louisiana, population 49,000.

Gina glanced at a gaggle of children waiting their turn at the top of a spiral slide. "How's summer school going?" she asked.

"Yesterday, I had one kid pop zits and another pop pills in the same class. The pecker-heads." Jo-Ellen chuckled and motioned toward the swings. "Look. Austin's fixing to jump."

Gina's stomach lurched as her oldest son bailed out of the swing, sailed through the air and landed feet first in a bed of pebbles. She turned to Jo-Ellen. "He scares the daylights out of me when he does that."

"I was flying." Austin picked himself up and dashed off to play.

"Just like a bird." Jo-Ellen flashed a big smile.

"Do you see Jesse anywhere?" Gina craned her neck and shielded her eyes from the sun.

Jo-Ellen pointed to a stationary fire truck made of wood and old tires. "He's racing to a fire." Jesse gripped the steering wheel and turned it round and round.

Relieved, Gina watched Austin join his brother. At the rear of the fire truck, a steel ladder hooked onto a set of monkey bars. Austin scrambled up the rungs, hung upside-down and made monkey noises.

"Your boys are cute," Jo-Ellen laughed.

"Thanks, but they can be ornery little devils at times."

"Can't we all," Jo-Ellen snickered. "Now tell me why you're so down in the dumps."

"It's all so crazy," Gina sighed.

Jo-Ellen adjusted her sunglasses and considered the younger woman. "You wanna talk about it?"

"It's Tuck." Gina kicked at the rocks.

"So I gathered. Y'all have a fight?"

"Sort of." Gina swung her leg over the bench, balled one fist, and rubbed it against her thigh. "He's been so damn cynical lately."

Jo-Ellen lifted an eyebrow.

"Okay," Gina chuckled. "More cynical than usual. You know what he did today when he got home from work? He tried to shoot Wynonna Sandford's dog with a pellet pistol. In broad daylight."

Jo-Ellen covered her mouth. "In officer housing?" She looked up at Gina, her voice muffled behind her hand. "Did Wynonna call the sky cops?"

"No. But Tuck said she went berserk."

Jo-Ellen leaned forward and peered over her sunglasses. "I don't blame her. Has Tuck lost his mind?"

"I'm beginning to wonder." Gina took a deep breath. "It all started about a week ago, when Tuck's best friend from pilot school got killed in England. Remember?"

Jo-Ellen gazed up at the sky, a deep somber blue this time of day. Rain clouds rolled in from the west. "Did Tuck ever call the widow back?"

"No. Jeri just called the one time. Whenever I bring it up, Tuck clams up. He puts on this big callous act. He's good at denial you know."

"Most men are. Sounds like he's hurtin' real bad."

Gina nodded. "I wish he'd talk to me. I hate it when he tries to act so tough and macho every time a pilot dies. I swear to God, I think the Air Force brainwashes these guys when they go through pilot school."

Jo-Ellen pulled off her glasses and stared up at the sky. "Smells like rain," she said quietly.

Gina raked a hand through her hair. "Tuck's ex-wife called about an hour ago. Sally's sending Tuck's teen-age daughter to live with us while she flits off to South America to save lost souls."

"How old is the girl?" Jo-Ellen scratched her ear.

"Sixteen."

"I take it you two don't get along?"

Gina shrugged.

Jo-Ellen studied her for a moment. "How long were they married?"

"A couple of months. Right after he got back from Vietnam. According to Tuck, Sally got religion and turned into a hippie. One night after work, he found their apartment lit up like a Catholic Church. Sally burned candles and incense in memory of all the Vietnamese kids she accused Tuck of killing with Napalm."

Jo-Ellen watched a dark-haired girl skip across the

planked wings of a wooden jet designed to look like the A-10 fighters assigned to the base. The child waved at someone and vanished inside the kid-size cockpit.

"Why did he marry her in the first place?" Jo-Ellen asked.

Gina dug her feet into the pebbles. "Blonde hair and big boobs, I guess. He never talks about it."

"What's your step-daughter's name?" Jo-Ellen gnawed at a chipped fingernail.

"Michelle." Gina folded her hands on her lap.

Jo-Ellen reached over and patted Gina on the hand. "If she stays long enough and enrolls for fall semester, she could end up in one of my classes."

"You don't know how lucky you and Buzz are. Not having to mess with step-children and ex-spouses."

Jo-Ellen slipped on her sunglasses. "Gina," Jo-Ellen's voice caught. "I have a confession to make. A long time ago, back when I was a size ten, I was married to a fighter pilot. He was killed in a plane crash."

Gina's throat clogged. She closed her eyes, trying to digest the news, her worse nightmare, besides loosing one of the kids. Finally, she looked up. "Why didn't you mention this when we first met?"

Jo-Ellen reached into her purse and pulled out a pack of cigarettes. "Men aren't the only ones good at denial." She shook one out of the pack, lit up, and took a long, deep drag. The smoke filled her lungs before she blew it out in tiny, gray smoke rings above her head.

The cigarette's red tip reminded Gina of a tiny afterburner.

"I'll tell you about it some other time." Jo-Ellen scissored the cigarette between her fingers. She flicked her ashes on the ground. "Look, I don't blame you for being upset about Tuck's daughter, but I think there's more to it than that."

Gina felt her mouth twitch, the way it did whenever she

was about to reveal some aspect of herself she usually kept private. She hesitated, then told Jo-Ellen about the string of wrong numbers and her fear of Tuck messing around.

Jo-Ellen puffed on her cigarette. "Look, Gina, I don't think you have anything to worry about. Tuck's crazy about you. You can see it in his eyes whenever he looks at you."

Before Gina could respond, Austin and Jesse ran up, panting.

"I'm thirsty," Jesse whined.

Gina stood up and caught their sweaty bodies. "Jo-Ellen, can we bum a ride home?"

"Under one condition." Jo-Ellen snuffed out her cigarette. "These crabby ol' boys better give me a hug first."

Tuck's 'Vette was gone when Jo-Ellen pulled her maroon New Yorker into the driveway. A brown Bronco with kids' fingerprints smudged on the windows was parked under the carport. The front door of the house was closed.

"I forgot my keys." Gina rubbed her arms to stay warm. Jo-Ellen had the air conditioner cranked up full blast to drown out the smoke from her current cigarette.

"I bet Dad left the door unlocked." Austin jumped out of the back seat to go check.

"I bet he did, too," Jo-Ellen chimed after him.

"I'm hungry," Jesse grumbled from the back seat.

"Hush, Jesse," Gina said, "You're not dying."

"If he didn't," Jo-Ellen said, "we'll leave Tuck a note, and y'all can come home with me." She flicked her ashes into the ashtray and twisted around to face Jesse. "As for you, young man, I've got a package of hot dogs in the freezer with your name on it. All I have to do is nuke 'em in the microwave."

Jesse cheered up instantly.

"You're a dear," Gina said.

Jo-Ellen motioned toward the house. "See, what did I tell you." Austin waved at them and disappeared into the house.

"Thanks for the lift." Gina gave Jo-Ellen a hug and

climbed out of the car. "C'mon, Jesse. Let's go see what we can scrounge up for supper."

"We'll do those hot dogs some other time." Jo-Ellen blew Jesse a kiss and backed down the driveway, a cigarette dangling from her mouth like a truck driver.

Gina had the boys fed and in bed by eight-thirty. She set their cereal bowls in the sink and went into the living room to wait for Tuck. Crouching in front of a small cabinet, she winced as her knees creaked and popped. The pain got worse every year. Running did wonders for her mind, but hell on the knees. She pulled out a shoe box filled with black and white photos of her parents' wedding. If she hadn't confiscated them after the divorce, her mother would have burned them. In the photos her mother was eighteen, shy and unsure of herself, with a virginal white veil trailing behind her. Gina's dad was twenty-one, a lanky airman from the base with a duffel bag full of dreams.

Gina stared into the photo, as if those distant young faces held all the answers. She put the pictures away and pulled herself up, coming face to face with a framed photo of her own wedding day at the chapel at Cannon Air Force Base.

Gina picked the photo up and rubbed it across her bottom to wipe off the dust. She was twenty-one, in an ivory chiffon dress with an empire waist, and Baby's breath intertwined in her pile of dark hair.

Gina's eyes drifted to Tuck, in his white mess jacket with captain's bars on each shoulder, his wings and medals over his heart. He was wearing a black cummerbund, a bow tie, and black slacks. His lips tweaked in a sly grin, like he prepared to sing: *Off we go, Gina, into the wild blue yonder...climbing high...into the sun...*

Gina set the picture back on the shelf and let her eyes sweep over the outdated architecture of the room, camouflaged by an eclectic mixture of Navajo rugs, limited-edition prints, shelves of books, and comfortable furniture. No telling how many other military families lived there before them.

Despite her and Tuck's vagabond existence, Gina was still a long way from Llano—and the shabby rent houses with their threadbare furniture and boxes of broken dreams—her father moved them in and out of every six months. Far away from the long nights when her mother kept Daddy's dinner warming in the oven because he was supposedly out peddling life insurance.

If her dad stayed in the Air Force, things might have turned out differently. If she hadn't found the panties, maybe her parents would still be married. If Tuck would open up more, instead of keeping everything inside. If...

She squared her shoulders and went to pour herself a glass of wine. She reached into the cupboard for a goblet and saw a note she somehow overlooked on the kitchen counter. The handwriting was neat and precise, like always:

Babe,

Wheaties called after you and the boys left. I'm meeting him at the club for a beer. Jeri Sweeney called from D.C. They buried Sweenedog in Arlington this morning. (What was left of him.) We'll talk about Michelle when I get home.

I love you,
Tuck

P.S. The ice cream's in the freezer for the boys. I'll stop by the shoppette on my way home and pick up some chocolate syrup for Jesse. Keep the doors locked and don't answer the phone.

Gina read the note twice. The angry woman in her wanted to lash out at him for keeping secrets. But the little girl in her cried.

Chapter Four

At Beauregard A.F.B. Officer's Club, the forty-seven-year-old wing commander stood ramrod-straight in a custom-tailored flight suit that contoured to his compact body. What Colonel Glen P. Dennison lacked in height, he made up for with a taut chest and flat stomach that put some of his junior pilots to shame.

He gripped Tuck's shoulder like old friends. "Tuck. How's it going?"

Hunched over a barstool, Tuck nursed a beer and waited for Wheaties to get back from the men's room. "Taking up space, using up oxygen, Sir," Tuck glanced sideways. "And you?"

The colonel released his grip. "Can't complain." His hard mouth barely opened when he spoke, giving Tuck the impression the colonel clenched his teeth all the time. "Let me buy you a beer," Dennison ogled the new bartender, a young Korean woman with shiny black hair down past her waist. The bartender operated the blender.

"Kim, Dear," Dennison called.

"Hang on a minute," the young woman yelled over the whir. "I have only two hands."

Colonel Dennison smoothed a hand over his salt-and-pepper hair, clipped short and neat. He winked at Tuck. "I think she's a tad frazzled her first night on the job."

Tuck thought, *Great, go for it Kim. Make the wing commander wait. Put God on hold.*

Cher sang *If I Could Turn Back Time* from the jukebox.

If Tuck could turn back time, he and Sweenedog would still be booger-eatin' lieutenants pumped full of Arizona sunshine and sappy ideology. Instead, he sat in the middle of swampy Louisiana with a full-bird colonel breathing down his neck.

Another typical Friday night at the O'Club swarmed with people and the sounds of clinking glasses, throaty laughter, and the warm buzz of a hundred conversations.

Colonel Dennison stuck a peppermint in his mouth and waited for the bartender to turn around. She turned off the blender and poured frozen margaritas into two giant goblets.

Tuck felt sorry for her when she picked up the margaritas and spun around. Her almond eyes froze at the sight of the wing commander's mocking expression.

"I'm sorry, Sir," she said. "I had no idea…" At that moment a second lieutenant sidled up to the bar, nodded to Colonel Dennison and Tuck, paid for the drinks and scrambled away with the two margaritas.

Colonel Dennison flashed Kim a haughty smile and placed a twenty-dollar bill on the counter. "Lieutenant Colonel Westerfield needs a Coors Light—make that a long neck—and I'll have the usual—scotch and water."

When Kim returned with the drinks, her small, delicate hands trembled as she fumbled with the cap on Tuck's beer.

"I'll get that," Tuck offered. He reached into a pocket on the right leg of his flight suit and fished out a handful of loose change.

She slid the beer toward him.

He dropped a few coins into a tall glass sitting on the bar for tips.

The bartender gave him a timid smile, then turned abruptly to Colonel Dennison, took the twenty, and rang up the register. She made every effort to avoid eye contact.

When she tried to give Colonel Dennison his change, he tucked the wad of bills into her hand in a flirtatious manner. "Keep the change, dear. It's your first day on the job."

Kim hurried off to wait on another customer.

"She reminds me of a girl I met in Bangkok—before I got shot down," Colonel Dennison remarked.

"How long were you in the Hilton?" Tuck referred to the time Dennison spent in a prison camp in North Vietnam.

"Almost a year. Not as long as some." Dennison paused to sip his drink, let his eyes dart around the room, then back to Tuck. "But long enough to know when times are tough, it pays to know who your friends are."

Tuck held his gaze. Dennison looked away first. Tuck took a swig of beer. His neck hurt.

Dennison leaned closer. "We're all on the same team, Tuck. We all want what's best for our careers."

"Yes, Sir," Tuck said. *I'll nod my head a few times and agree with everything you say, Sir. Then you'll get off my back.*

The colonel's thin lips slid back in an alligator grin. "You're a good man, Tuck." He lowered his voice. "Keep your slate clean and I'll see you get that squadron. Forget about the other night and I'll forget about Killer's bachelor party." Dennison straightened and glanced around the room. "By the way, how are Gina and the kids?"

Dennison struck a nerve.

"They're fine." Tuck focused on his Coors Light.

Dennison nodded, then turned and waved to his wife in the doorway. Some people might describe Lois Dennison as mousy, but Tuck thought just plain.

"If you'll excuse me," Dennison said. "I see my bride." He slapped Tuck affectionately on the back and gestured toward the unopened bottle on the counter. "Drink up, Tuck. Your beer's gettin' warm."

Tuck blew the air out of his lungs after Dennison left. Consorting with the enemy took work. He swiveled around on the stool and watched Dennison weave his way through the crowd. The motto he and Sweenedog learned back in pilot school came to mind: *A warrior never shows weakness to his enemy. He moves from a position of strength.*

Tuck surveyed the bar packed with more than four dozen pilots from all three of Beauregard's tactical fighter squadrons. Wives to girlfriends to a cluster of skimpily dressed wannabes who drove out from Bolton in hopes to snag a pilot.

Near the back of the bar, a group of ground pounders, what the pilots called nonflying types in combat support units, such as civil engineering and administration, sat at their own tables. In an officer's club on a fighter base, fighter pilots ruled.

Tuck slid off the barstool and walked to the jukebox. He plugged in a quarter, made his selection, and leaned against it as Bob Seger came on, growling *Like a Rock.*

Tuck listened to the music while an old conversation with Sweenedog a decade ago popped into his head

"I got my dream assignment," he bragged over the phone to Sweenedog. Tuck flew F-llls out of Cannon AFB, New Mexico, and Sweeney, out of Mountain Home, Idaho. *"I'm going to Davis-Monthan to fly A-10s."*

"Single seat," Sweenedog crowed. " I'd give my left nut to fly single seat."

But for some reason, Sweenedog never got out of the F-lll program. He never tasted the freedom of flying without a navigator.

Leaning against the jukebox, Tuck's eyes lingered on a handful of old heads like himself, huddled around a nearby table telling jokes. Though Tuck's age and rank, none had flown in combat. While Tuck flew F-4s in Southeast Asia, these men were instructor pilots stationed back in the states in Air Training Command.

Until a week ago, Tuck begrudged this small group of pilots. He glanced at the wing commander and his wife, seated nearby. *Combat didn't make you a better person, a better pilot,* Tuck decided. *It just made you older. If you were lucky to live through it. Regardless of whether you fought in an actual war or trained for war, to dodge death took work.*

"Hey, play something we can dance to," somebody shouted over the hubbub of the crowd.

A blue-eyed blonde in a white mini-skirt sauntered up to the jukebox and flashed Tuck a big white smile. "Mind if I play some Tracy Chapman," she purred.

She reminded Tuck of a young Farrah Fawcett. He moved out of the way. "Who's Tracy Chapman?"

"You know. She sings *Fast Car*?"

Tuck watched her feed the jukebox quarters.

He worked his way back to the bar when a snappy tune came on. Something about getting in a fast car.

Tuck felt like a dinosaur.

He ordered a glass of ice water, leaned against the bar, and crunched ice. He thought about Gina and the long, empty months before they met each other, when he first got to Cannon O'Club, all dick and no forehead, and woke up many mornings with a hangover and a different woman in his bed. Take Jane, the attorney's wife from town with the topknot, Lincoln Town Car and one too many face lifts.

"We have an arrangement," Jane breathed Tuck's first night on base.

Or Trixie, the snake-eyed blonde in the yellow Corvette and Loretta, the waitress who hunted for a husband, a ticket out of town and a daddy for her four kids.

Tuck hated to think where he would have ended up if he hadn't met Gina in 1974.

Tuck and a navigator from Memphis drove to Portales,

a college town twenty miles south of Cannon, on the New Mexico/Texas border to check out some of the nightspots. When they walked into the bar, Tuck found himself drawn to the fresh-faced brunette with the high cheek bones and freckles seated with another girl at a corner table.

After ordering their drinks, Tuck and his friend swaggered over to the girls' table, and the navigator started doing Elvis impersonations. The other girl squealed with delight and invited the navigator to sit down. Bubbly and cute, she wore enough make-up and hair spray to last through the night, in case she got lucky.

The brunette rolled her eyes at Tuck. "Another bullshit artist at work. Pull up a stump." Gina Loyd, a psychology major, was quiet and unassuming, with discerning hazel eyes that stared straight into Tuck's head like she could read his mind.

Tuck blushed and felt vulnerable. To regain his balance, he heaped on the war stories to impress her. But the more he talked, the deeper her eyes penetrated into his head. Out of frustration, he drank more than he should to wash down the feeling her eyes had blown his cover.

By the end of the evening, every time he belched, his head bobbed like those little purple cows people use to stick on their dashboards back in the seventies. "Can I call you?" he hiccupped, his speech slurred.

She scribbled her phone number. "When you're sober."

When Tuck called her the next day to apologize, she chuckled, "Ah, the world's greatest fighter pilot."

They got married a few months later.

A cacophony of voices jarred Tuck back to the present. Over the blare of Survivor singing *Eye of the Tiger,* several young women kicked up their legs in an impromptu can-can on the dance floor. One voice stood out.

"Ladies, let the master of fun show you how it's done," Wheaties shouted, back from the latrine.

The big lieutenant with a red flattop pulled off his

squadron scarf and tied it Samurai-style around his head. At six-foot-two, with a thick neck and square chin, Wheaties looked like a Dallas Cowboys linebacker. He hooked into the center of the dance line and kicked up his legs, showing off for the crowd.

A few seconds later, he untangled himself from the dance line and howled, "Okay hog drivers," to his fellow A-10 pilots, who affectionately called their jets warthogs or hogs for short. Lifting a shot glass of tequila in a toast, he crowed, "It's live to drive and drive to live," before slamming it down. The crowd roared in approval.

When he finished, he set his glass down, wiped his mouth with the back of his hand, and hooked back into the dance line.

A leathery woman in polyester slacks and gold slippers sidled up next to Tuck. Her gray hair butchered just above her ears, she smelled of cigarettes and booze. Liver spots covered her gnarled fingers adorned with dinner rings. *Some retired colonel's wife*, Tuck figured.

"Who the blazes is that obnoxious lieutenant?" she slurred in a scratchy voice. Her lipstick bled into the craggy skin around her mouth, an angry red slash on a small, grouchy face.

Tuck looked at her and grinned. "That's Wheaties Ma'am. He'll be Chief of Staff of the Air Force one day."

"Smart-ass," she sneered, picked up her drink and pattered off.

She reminded Tuck of his mother. Disenchanted and angry at the world. *Why can't you be a lawyer, Tucker, or a doctor, or a businessman like your daddy*? His mother had lashed into him on his sixteenth birthday. *I don't know where you got this dream to fly. Nobody else in the family flies. I don't care what your granny says. She's not paying for your flying lessons.*

Tuck's attention went back to the dance floor. The can-can dancers broke formation, most of them heading for the ladies' room. Wheaties stood in the middle of the dance floor, rubbed his hand back and forth over his head of red bristles.

The lieutenant spotted Tuck and headed his way.

"There you are, Sir," Wheaties grinned. "I got sidetracked by some wild women." His freckled face glistened with sweat, and perspiration stains left white circles under the armpits of his flight suit.

"Check six," Tuck warned, pointing with his chin. "Dennison just gave you the evil eye."

"Shoot. Guess I better start behavin' myself," Wheaties said.

"Not unless you want a one-way ticket to Kansas."

Wheaties laughed and then turned serious. "Uncle Mathew would never forgive me if I got kicked out of the Air Force. That old coot helped Mama raise me after Daddy got killed."

Tuck took a sip of ice water. "How did it happen?"

"Sprayin' tomatoes. His engine quit. He didn't pull up in time. Got wrapped around some high wires."

"How old were you?" Tuck set his water down.

"Nine." Wheaties paused and looked out over the crowd. "Uncle Matthew felt responsible, being the oldest. He started flying crop dusters after he came back from Southeast Asia. Then Daddy got his pilot's license, and they became business partners.

"Mama swore Uncle Matthew was already messed up in the noggin' when Daddy died." Wheaties rapped his knuckles against his head. "She always blamed it on Vietnam. Said Uncle Matthew was never the same after the war."

Tuck nodded and Wheaties fell silent, lost in thought.

A short time later Wheaties elbowed Tuck in the ribs. "Sir, I think Major Sandford's wife is giving you a dirty look."

Wynonna Sandford stood at the other end of the L-shaped bar glaring at Tuck.

"She claims I tried to murder her mutt today." Tuck kept a straight face.

"What do you have against her dog?" Wheaties glanced at Wynonna then back at Tuck.

"What do they have against pilots?" Tuck rattled the ice in his glass.

"Well, don't look now," Wheaties said. "Here comes trouble."

Wynonna strutted past them, her nose in the air.

"Evenin', Mrs. Sandford," Wheaties said.

She stopped and glared at Wheaties' name tag. "Lieutenant," she batted her lashes. "I'm sure you're a nice boy, so take my advice. Stay away from people like Colonel Westerfield." She scowled at Tuck.

"But Ma'am," Wheaties said, "Colonel Westerfield taught me everything I know."

"Why thank you, lieutenant," Tuck beamed.

"Fly-boys," she scoffed, tossing her nose in the air at Tuck. "Y'all get preferential treatment. Must be nice havin' the wing commander chauffeur you around in the white top."

Tuck's stomach dropped. "I don't know what you're talking about," he lied.

"You know what I'm talkin' about, Mister High-and-Mighty," she sniffed. "I saw you get out of Colonel Dennison's staff car the other night when I let Baby out front to potty. I told Big Sandy. He says you're brown nosin' for a promotion."

"You tell Big Sandy to mind his own business."

Wynonna used a fingernail to scoop out a cherry floating in her cocktail. "You know," she chomped, so close to Tuck he smelled the cherry and whiskey on her breath. "Big Sandy's not gonna take too kindly 'bout you tryin' to shoot Baby when he gets back from Langely. He don't like dog killers."

"I bet he won't appreciate getting written up for breaking the base leash law, either."

Wynonna turned in a huff and clip-clopped off, parked herself at the ground pounders' table.

Wheaties scratched his head, bewildered. "Sir? What did

she mean by that? You getting dropped off in the white top?"

"Nothing, Wheaties." Tuck rolled his eyes. "She's a crackpot."

Wheaties shrugged. "Hey, I'm just a dumb lieutenant from Kansas trying to serve God and my country the best way I know how."

Tuck slapped him on the back. "Time to recycle some beer."

When Tuck walked out of the men's room, Captain Linda Garrett, the wing commander's executive officer, stopped cold in front of Tuck at the end of the narrow hallway.

Tuck's body blocked her path to the ladies' room.

She stiffened.

Tuck smiled down at her.

"Isn't it dangerous kissing up to the boss when the wife is around?" He motioned back toward the casual bar, where the music drifted out into the hall. "Lois is with him. Just thought I'd warn you."

Nostrils flaring, the captain said in a low voice, "If you'll excuse me, Sir."

Tuck yawned. "With all due respect, Captain, you've got a husband and kids. What's in this for you?"

The captain looked away.

Tuck glanced at his watch. "Well, give my regards to the colonel." He stepped aside to let her pass, turned and strolled down the hall.

He rounded the corner, passed her husband, Major Hank Garrett, a pilot in another squadron, parked on one of the plush sofas out in the entryway.

"Evening, Hank."

The amicable-looking major looked up and waved.

Wheaties laughed about something when Tuck rejoined him at the bar.

A short, stocky captain walked past and punched Wheaties in the arm. "Yo, Wheaties, how's it going, man?" He nodded respectively at Tuck. “Sir.”

Wheaties looked up and grinned. “So Killer, how’s married life treating you?”

Killer leaned closer. “I’m not pussy whipped if that’s what you mean.” He winked at Tuck and put his hands on Wheaties’ shoulders. “Not like the rest of you ol’ married guys.”

Wheaties burst out laughing.

Killer strutted off, joined a cute brunette at a nearby table. His new wife hooked a possessive arm around him and pulled him down next to her. Killer sat with a sheepish look on his face while his wife flashed a fat diamond ring around at their friends.

“Poor guy,” Wheaties mused. He swiveled around on the barstool. “Let’s hear some war stories from an old head, Sir.”

For the next fifteen minutes, Tuck’s right hand trailed his left in simulated flight. His mission was to give guidance to a young pilot like Wheaties so that one day, when he takes off or lands, or dives low over the bombing range to fire the A-l0s Gatlin gun, or flies in close formation, he won’t kill himself.

When someone cranked up the volume on the music, Tuck decided to head home.

He set his glass on the bar to leave. “Stop by the house for a beer sometime,” he told Wheaties.

On his way out, Tuck bumped into the blue-eyed blonde he encountered at the jukebox.

“Leaving so soon?” she asked. “The party’s just getting started.”

“I gotta pick up some chocolate syrup and go kill fire ants.”

“Chocolate syrup kills fire ants?”

“No. Poison kills fire ants.”

“Oh but,” she scrunched up her face, “it’s dark outside.”

“That’s why God invented flashlights.”

At the top of the steps outside the main entrance of the

club, Tuck put on his flight cap when two men approached, one tall and thin, in dress blues, a non-rated type, the other a big, burly pilot smacking on a piece of gum.

Tuck grimaced, then put on a big grin.

"Well, well, well, if it isn't Laurel and Hardy." He rattled his car keys in the air. "We meet again."

The man chewing gum pulled off his flight cap, revealed a hawkish nose and a plumage of white hair slicked back like a bald eagle. A vein in the middle of his forehead pulsed to life, a clear indicator of a rise in his blood pressure. "Knock it off, Tuck," said Lieutenant Colonel Leland "Bull" Spitz, the beefy squadron commander of the 428th.

Ah, Bull Spitz, Tuck thought. *Another old head who managed to dodge combat during the Vietnam War.*

Tuck turned his attention to the other man, Lieutenant Colonel Vic Rollings, Beauregard's Chief of Personnel and resident UFO geek who scanned the nightly skies from his rooftop on base.

"Tracked down any extraterrestrials lately?" Tuck asked the confirmed bachelor, a fussy sort of fellow who lived next door to Bull Spitz on Bayou Way.

Vic Rollings pushed up his black Air Force-issue glasses and glowered at Tuck. "Another skeptic," he snipped, sweeping his eyes upward.

The 428th commander frowned at Tuck, then turned to Rollings. "Go on, Vic, I'll catch up with you in a minute. Get me a gin and tonic, will you?"

"Chummy, aren't we?" Tuck said after the Chief of Personnel grumbled up the steps and went inside. "Since when did you start slumming with ground pounders*?*"

Bull Spitz stiffened, his blue eyes frozen on Tuck. "Come on, Westerfield, let's all get along and play by the rules. We have an important friend to protect, not to mention our own careers."

"Sure thing, Bull, anything you say." Tuck played with

his keys. "But first, I want you to stop calling my house to check up on me. Christ man, you act like you're paranoid." Tuck twirled his key ring on his finger. "Some might even consider your phone calls a form of sexual harassment."

"This isn't funny, Tuck." Spitz stopped chewing his gum.

"Who's laughing? By the way, if you ever threaten my wife or kids, I'll go straight to the Office of Special Investigations. Comprende?"

Spitz snarled something under his breath and stomped off.

Tuck headed out to the parking lot. The back of his neck throbbed overtime.

Chapter Five

On Monday, July 2 at 0700, Major Big Sandy Sandford rolled out of bed and pulled on a pair of dark-blue slacks draped over a chair. He sucked in his gut and tucked in his light-blue, short-sleeve shirt. In a hurry to see Colonel Dennison, he shaved the night before and went to lace up his scuffed wing tips and clip on a dark blue tie. For Beauregard's Chief of Civil Engineering, the Air Force uniform was an evil necessity. The six-foot-three Texas Aggie with the ruddy complexion lived for the weekends and his grubby jeans and cowboy boots.

On his way out, he bent over the bed to tell his wife goodbye.

"I'm leaving, Schnookums." He nudged her on the derriere.

Wynonna Sandford sprawled on top of the covers in a filmy nightgown. A sleeping mask shielded her eyes.

She stirred and yawned thickly. "Okay, Daddy."

Big Sandy grabbed his pouch of chewing tobacco off the night stand. "Good luck in Rose Glen. What time you going?"

"After lunch." She peeked at him out of one side of her

mask. "It'll take me all morning to fill out orders and load the Suburban."

"I'll call you and let you know how it goes with Colonel Dennison."

Wynonna reached protectively for her dog.

At the bedroom door, Big Sandy turned. "Don't you worry, boy. After today, that pilot's lunch meat."

The Pekingese glanced up, blinked at him, then curled closer to his mistress.

At 0720 Big Sandy climbed into his lumbering yellow Cadillac, a veritable land yacht that had seen better days, backed out of the driveway in one swift move, and flipped a birdie in the direction of the house next door.

"Damn fighter pilot," he snarled.

A few minutes later, he wheeled into an open slot in front of Wing Headquarters and vaulted up the stairs to the second floor.

Colonel Dennison frowned behind a massive cherry desk when Big Sandy walked in. "Yes, Sandford?" His eyebrows rose sharply. "What was so important you couldn't wait for an appointment?"

"Well, Sir," Big Sandy choked, "It's about my dog, Sir."

Fifteen minutes later, the door to Colonel Dennison's office swung open and Big Sandy swaggered out. He trotted down the stairs and Captain Linda Garrett came toward him.

"Good morning, Captain," Big Sandy beamed. He waited for the flat-chested blonde to salute him.

"Good morning, Major." Her navy pumps tapped up each step when she breezed past him without a salute.

Big Sandy shrugged. Nothing could spoil his mood, not even this conceited captain. Married to a pilot, she worked as a gofer for another pilot, the commanding officer.

And pilots got all the glory.

Take his older brother, Clem, the handsome one. Clem

the future Air Force pilot and favorite son. Clem was dead, but Big Sandy still tried to fill his shoes, and got himself washed out of pilot school in the process.

A feeling of failure came over him, but he quickly dismissed it and moved for the door.

Out in the Caddy, Big Sandy pinched off a wad of chewing tobacco, plugged it under his bottom lip, then revved up the engine. The meeting upstairs had gone well even though Colonel Dennison told him to get a haircut. As Big Sandy pulled away from the curb, and headed for his own office down the street, his mouth contorted into a jagged, satisfied grin.

***

After Major Sandford left, Colonel Dennison shut the door to his office and moved to the large window behind his desk. He glanced at the parking lot below, then picked up the phone and punched in a private number to the 428th.

Lieutenant Colonel Bull Spitz came on the line. The wing commander kept his voice low. "Looks like you might be right about Westerfield. He's not a team player."

"Did he talk?"

"Not yet," Colonel Dennison said. "But I don't trust him. Something's come up."

"What do you mean?"

"Nothing we can't handle. Let's just say I've been handed a nice piece of rope to hang Westerfield. At least slow him down for a while, in case he feels the urge to talk."

"And what's that?"

"That slime-ball Sandford came here with some asinine story about Westerfield threatening to shoot his wife's dog."

The squadron commander laughed.

Dennison ignored him. "Westerfield's car isn't in the parking lot. I assume he's flying?"

"As we speak, Sir."

“Call Westerfield back from the range. Tell him to get his butt over here PRONTO!”

“Gladly, Sir,” Bull Spitz complied with relish.

***

The first green A-10 Thunderbolt II made a pass over the target area, then pulled up and away. A second plane dived in low over the bombing range, skimmed above the tree line, and dropped a bomb, then pulled up and away. Its engines screeched like a vacuum sweeper. The plane banked effortlessly to the right and followed its leader.

“Bull's-eye number two,” boomed the voice of the range officer over the radios of both planes.

In the lead plane, flight leader Lieutenant Colonel Tuck Westerfield glanced swiftly over his left shoulder at the hunk of twisted metal smoldering 3,000 feet below. He radioed his wingman. “Congratulations, Tony. At least one of us had a good day on the range.”

“Can't even tell it was an old fire truck,” boasted Captain Tony Grimes in his warm southern drawl, broadcasting loud and clear over Tuck's radio. “Guess I got lucky.” Tony was ten years younger than Tuck and the only black pilot in the 428th.

“Lucky my ass,” Tuck chuckled. He re-positioned his A-10 for another pass over the bombing range.

Tuck had been the 428th's Top Gun three months in a row, but this morning his bombing sucked.

Both pilots dropped practice bombs on the range since 0745 that morning.

At 0755, the voice of a female air traffic controller broke in over the radio:

“This is Beauregard Tower. Lieutenant Colonel Westerfield and Captain Grimes, return to base immediately. Wing commander's orders.”

"Shit," Tuck hissed.

He zoomed along in his jet toward the drop zone, concentrated on his gun sight and the approaching target. An old yellow school bus grew bigger by the second.

Return to base immediately, the controller said. Tuck couldn't ignore the direct order.

He pulled back on the stick and banked to the right, away from the bombing range. The implications began to sink in. A plane had gone down. Colonel Dennison needed Tuck to do an in-house investigation before the official Air Force accident board convened. There could be no other explanation for such an urgent message, outside of war. Tuck didn't care any more about his lousy bomb scores.

***

At Wing Headquarters, Tuck slipped in through a side door and stopped by his office. The young safety clerk looked up in surprise.

"I thought you were flying, boss," said Sergeant Juan Duran before he rattled off a string of messages, none of them urgent. Tuck took a deep breath and went upstairs.

"I'm disappointed in you, Tucker," the wing commander said. "We agreed to work as a team."

Tuck stared into the dark, unyielding eyes and the hard mouth. The colonel wore blues, instead of a flight suit. His freshly starched shirt appeared as stiff as the man himself.

Dennison leaned forward. "Something has been called to my attention that clearly demonstrates bad judgment on your part."

Tuck narrowed his eyes in confusion. "Sir?"

"I simply cannot tolerate my Chief of Safety threatening a neighbor's dog with a gun."

Tuck almost laughed, then thought better of it.

He could clear this up in no time and get back to the

office. Settle with Big Sandy later over a beer. Then patch things up with Wynonna, who obviously knew as much about guns as how to dress tastefully.

"Sir, the gun in question was a pellet pistol," Tuck explained.

"A pellet pistol*?*"

"Yes, Sir."

Colonel Dennison leaned back in his chair. "And you were only trying to scare the dog? Is that it?"

"Yes, Sir. You see, Sir, the Sandfords have broken the leash law for two years. I can assure you, Sir, it's not a very pleasant task to come home from work everyday and scrape dog poop off your kids' tennis shoes."

Colonel Dennison's eyebrows shot up. "Tell me something, Tuck. Why didn't you simply report them to the housing office?"

"To tell you the truth, Sir, I figured it would cause more problems than it was worth." Tuck scratched his chin. "Seems me and Major Sandford haven't gotten along from the beginning. He has a prejudice against pilots."

Dennison sprang forward in his chair. "He's a damn ground pounder, Tuck. What did you expect?"

Tuck swallowed.

Dennison cleared his throat. "I hope you don't think a pellet pistol makes it all right, or justifies your actions?"

"No, Sir. The unfortunate incident won't happen again, Sir. You have my word."

Dennison took a deep breath. "The fact that the gun was a pellet pistol is immaterial. Don't you see, the whole thing just doesn't look good."

Tuck clenched his jaws and willed himself to remain calm. *Never show weakness to your enemy.*

Dennison knotted his brows. "You do understand that I'll have to exercise some form of punishment?"

Tuck felt lightheaded. The Instant Breakfast gulped early

that morning curdled in his belly and gave him indigestion.

"The only recourse I have is to ground you."

Tuck cringed at the injustice. He felt like he'd been kicked in the gut.

"Your grounding is only temporary," Dennison continued. "Time away from the cockpit might do you some good. Give you time to reflect on a few things, say, your name on the squadron commander's list? You made a bad judgment call. An easy mistake. Every officer will make at least one during his career."

Dennison referred to himself, Tuck knew. Yet instead of looking away, the wing commander's steely-eyed gaze challenged Tuck with a silent warning: *Don't even think about reporting me. I am above reproach.* "That's better than an Article 15. We all know what that would do to your career."

Tuck half listened. All his instincts told him to fight back, but something deep within forced him to sit tight. For nineteen years, Lieutenant Colonel Tucker F. Westerfield was taught, no, conditioned, you never talked back to a superior officer. To argue speaks of insubordination and it didn't matter if you were right or wrong. The military wasn't a democracy; you didn't get a vote—you followed orders.

"If anybody inquires about why you're not flying, tell them you threw your back out," Dennison said. "We'll make it look like a medical. Save you the embarrassment. By the way, don't stray too far from the phone today. Colonel Spitz will make you an appointment with the flight surgeon. Sorry, Tuck. Even I can't authorize a medical." Dennison rose to dismiss him. "That's all I've got. You're free to go."

Tuck gathered himself up and left. Outside, in the sticky-swamp of morning, he felt anything but free.

In the parking lot, like a good steed, sat his white Corvette to spirit him away.

***

After Tuck left his office, the wing commander called his own boss, a three-star general halfway across the country who happened to be his brother-in-law.

Lieutenant General Marshall Honeycutt's gruff voice came on the line. "Hello, Glen. How's my sister?"

"Lois is fine, Marsh. She's feeling much better since her surgery."

"She better be fine," snapped the three-star. "Now what have you got to report?"

"Our little problem is taken care of. Not to worry, Marsh. I've got Westerfield eating out of my hand." He didn't elaborate.

"I sure as hell hope so." The general raised his voice. "I can't keep bailing you out."

Dennison bit his tongue. "Yes, Sir." His voice thick with resentment.

"What about Sergeant Chambers. The young sky cop on patrol that night?"

"He won't spill his guts. He has a family to support."

"Our ass is grass if you're wrong," Honeycutt growled.

Dennison took a deep breath and measured his words carefully. "Not to worry, Marsh. I've got everything under control."

"What about the young lady?"

"Everything's been taken care of."

"And her husband?"

"Doesn't suspect a thing."

"I strongly urge you to keep that young man happy. Start working on his next assignment. I believe he wants F-16s?"

Dennison hesitated. "Certainly, General."

"And you're positive Bull Spitz won't present any problems or his sidekick, Rollings?"

"I told you, Spitz is one of us, and Rollings, well, he may be a space cadet but he's a team player."

"God help the Air Force," the general barked. "Do us all

a favor, Glen. Keep your pecker in your pants."

Glen Dennison stood at the window a long time, staring with contempt at the lush golf course across the street. Retirees in Bermuda shorts and gimmee caps wheeled around in golf carts. When they weren't playing golf, they hung around the clubhouse, scratched their balls, tried to remember somebody's name, and hunted for a place to pee.

In Dennison's eyes, these retired officers and NCOs were nobodies now, merely listed as Mr. This or Mr. That in their church directories. Or they shuffled along with their wives at the commissary, pushed their loaded-down shopping carts and jammed up the aisles.

Dennison grimaced. The thought of retirement terrified him.

Turning from the window, he reflected on the unique relationship he shared with his boss. They met at the Air Force Academy, when Marshall Honeycutt, an upperclassman, befriended Dennison, a first-year cadet.

Honeycutt became Glen Dennison's mentor and closest friend, close enough that the upperclassman introduced the young cadet to his bookish kid sister, Lois. "Be good to her, Glen," Marshall advised. "That's all I'm going to say."

And he was until he came back from Vietnam.

Vietnam. Dennison clenched his jaw as he recalled the day when Marshall Honeycutt screwed up, and changed their relationship forever.

They both flew F-4s out of Ubon when a pair of MIG-21s jumped them. Honeycutt was a major, the flight lead on the ill-fated mission, when one of the MIGs locked onto his tail and he couldn't shake it. Within seconds, the MIG lobbed a missile up the tail end of his Phantom, and the younger Captain Dennison swept in and shot it down, saving Honeycutt's life.

For reasons Honeycutt could never explain, he bugged out, leaving Dennison to fight the other MIG alone. Dennison ended up getting shot down, and he spent the remainder of the war in a prison camp.

Ever since, Marshall Honeycutt felt guilty for deserting his wingman, his kid sister's husband. When the POWs came home in 1973, the first thing newly promoted Colonel Marsh Honeycutt said to a gaunt Dennison as he stepped off the plane at Clark AFB in the Philippines was, "I owe you one, partner."

Colonel Glen Dennison had been Beauregard's wing commander for almost two years. He deserved to make general. Dennison knew his brother-in-law would take care of him. Marsh owed him.

Chapter Six

Gina toweled herself off and trudged across the room to the dresser. She yawned and slipped on her bra and panties. Tuck's alarm had gone off at four that morning; she never went back to sleep.

On those mornings before dawn when Tuck had to fly, Gina lay in bed with one eye propped open and listened to him move around in the dark. She memorized the noises he made—the zipper on his flight suit, the squeak in his boots, the clink of metal as he gathered up his loose change and dog tags from the night stand and dropped them into his pocket. Plus a 1935 silver dollar he never flew without.

"Be careful," she said when he bent over to kiss her goodbye.

"Roger," he replied before peeking in on the boys.

Later, when she could no longer hear the rumble of the Corvette going down the street, she lay wide awake and wondered if today Tuck wouldn't come home. After fifteen years, she never got use to it.

Rummaging through the dresser, she pulled out a T-shirt stamped with Tuck's old squadron emblem from Alaska. Despite the harsh winters, which could dip to sixty below, Alaska held a special place to Gina because she gave birth there. Yet one bad memory spoiled things to remind her that no matter how much life insurance a person had, it was never enough.

Gina stared at the shirt, and recalled the middle-of-the-night phone call in late March, when barely two months pregnant with Jesse. The Command Post called to speak to Tuck. Ten minutes later, he left the house in his dress blues and parka, on his way to face the wife of one of his squadron buddies and tell her the grim news, the man of her dreams, the father of her children, had just crashed his jet into a mountain in Norway. The job fell to Tuck, the acting commander in charge of the pilots who didn't deploy.

He never talked about what happened that night, but Gina could see in his eyes the next day the shattered face of a woman too young to be a widow and too old to start over. When Gina tried to inquire about the couple's four children, and how they were holding up, Tuck bolted for the shower.

Gina stuffed the shirt back into the drawer, grabbed one of Tuck's white T-shirts, and buried her face in it. She breathed in the familiar scents of warm skin and aftershave, the good smells she couldn't wash out of his clothes for some reason. Tuck's scent clung for dear life onto every thread of fabric.

After putting on the shirt, she pulled on a clean pair of cut-offs and zipped them up when the phone rang. Ignoring it, she went to check on the boys.

"Mom, the phone's ringing," Austin said when she walked into his room. The boys played Ninja Turtles. Austin started to get up.

Gina stopped him. "We'll let the answering machine take it."

"In case it's another wrong number?"

Austin's question caught her off guard. Although the phone calls had tapered off, Gina screened her calls during the day when Tuck was at work. Yesterday, when she came back from chapel, she asked Tuck if he'd been seeing another woman. With a weary look in his eyes, he shook his head no, kissed her on the forehead, and went to go shave.

"Get the phone, Mommy," Jesse babbled, sitting splay-legged on the floor next to Austin.

"We can't," Austin told him. "Some dummy keeps calling our house and it's getting Mom upset. Right, Mom?"

Gina stood in the doorway, and listened for the answering machine out in the kitchen. She marveled at her oldest son's insight, yet sometimes it annoyed her.

"Yeah," Gina shrugged, using the same vague tone of voice Tuck used on her sometimes. A tidbit of information here, and a tidbit there, leaving Gina's imagination to fill in the blanks. Didn't Tuck know what he didn't say worried her? But a military wife was left in the dark about a lot of things, secret airplanes and covert operations like the raid on Panama the previous December, and for now, Gina reconciled herself to this one disparaging fact. Some things a military wife would never know about. Maybe this was one of them.

The phone stopped ringing and the answering machine kicked on.

"Hi Hon—" Jo-Ellen's down-homey voice reverberated throughout the house.

Gina padded into the kitchen and plucked up the phone.

"Screening your calls?" chuckled Jo-Ellen.

Gina heard the click of Jo-Ellen's cigarette lighter. "What's up?" Gina yawned.

"Buzz just called me at work. He said there's this crazy rumor going around that Tuck's grounded?"

"Tuck's not grounded? He left here at the crack of dawn." She started to feel uneasy and glanced at the clock on the microwave, ten-thirty, and half the morning shot. A pile of laundry waited by the washer.

Jo-Ellen took a drag on her cigarette. "I know Honey, but

half the troops over in maintenance and civil engineering think he is. They're saying it's because he tried to kill the Sandford's dog."

Gina rinsed a cereal bowl and stacked it in the dishwasher. "That's ridiculous."

"To make matters worse, somebody in civil engineering got their wires crossed and said Tuck tried to shoot Major Sandford."

"Maybe he should have." Gina slammed the dishwasher door.

"Mommy," Jesse yelled from the hallway bathroom.

"Look, Jo. I gotta go. Jesse's on the pot."

"Call me when you hear something."

Dazed, Gina hung up the phone. She didn't know what to think. Crazy rumors always floated around on base. She never paid much attention to them. But this one rumor she couldn't ignore.

"I'll be right there," she hollered to Jesse then dialed Tuck's number at headquarters. Sergeant Duran came on the line.

"Hi, Juan." She stopped to clear her throat. "Is my husband around by any chance?"

"No, Ma'am, he left here awhile ago for a meeting with Colonel Dennison. Anything I can help you with?"

"Did he fly today?"

"I'm not really sure. But he did show up here at the office a lot earlier than I expected."

Gina drummed the eraser end of a pencil on the counter. "What time was that?"

"Sometime after 0-nine-hundred, Ma'am."

Nine o'clock? she thought, puzzled. *Tuck and Tony should still be airborne*.

"Shall I have him call you when he comes back?"

"Sure." She hung up.

Jesse wobbled down the hall looking for her. His shorts

and trainer pants bunched up around his ankles. “It’s messy,” he said.

“I’m sorry, sweet boy.” She scurried him off to the bathroom. “Mommy’s proud of you for using the big-boy potty.” They passed by Austin’s room.

Gina steered Jesse into the bathroom.

“Jesse’s been calling you.” Austin came up behind her and leaned against the door. “I’m glad Michelle’s coming, even if you’re not.”

“Me too,” Jesse piped while Gina plopped him on the toilet seat.

She sat on the edge of the tub and looked at Austin. “Honey, I didn’t say I wasn’t glad…”

“But you and Daddy were fighting.” Austin fidgeted with something in his hand. “You called Daddy a bad word.”

“Affhole,” Jesse chortled on cue.

Gina cringed, feeling guilty. Not only was her chubby-cheeked three-year-old learning to cuss like a sailor from his military father, but from his church-going mother, too.

“I know I did,” she sighed, eyes on her feet. “I’m sorry.” She gave Austin a hopeful smile and turned back to Jesse.

“All done, young man?” She stood up to help him.

The phone rang. She poked her head into the hall. The answering machine kicked on.

“Tuck. Bull Spitz.”

Gina bristled at the man’s arrogance.

“It’s ten forty-five. I assume since you’re not in your office, you’ve gone home for an early lunch. Meet me at the hospital at thirteen hundred hours. You’ve got an appointment with the flight surgeon.”

Gina’s lips quivered when she stooped to pull up Jesse’s training pants. One look into his big green eyes and Gina broke down. She dropped down on the edge of the tub and turned away so the boys wouldn’t see her cry. First the call from Jo-Ellen. Now this. What did it all mean?

"You sad, Mommy?" Jesse reached up to flush the toilet.

"Run along," she croaked behind a veil of hair in her face.

Austin snuggled beside her and locked his arms around her waist. "Was that one of those wrong numbers?" He sounded as bewildered as she felt.

A second later Jesse skipped off down the hall. "Affhole, affhole, affhole," he sang at the top of his lungs.

***

The white Corvette sped down the two-lane road, twenty-five miles south of Bolton and a world away from the air base. Something beyond Tuck's control seemed to pull him down the road, flanked on both sides by an ancient forest of towering white pines.

Ahead, the forest gave way to a huge, two-story Tudor, back in a clearing. A split-rail fence ran the length of the property. Horses grazed in a nearby pasture. But the majority of dwellings Tuck passed were either ramshackle houses with broken-down cars in the front yards or mobile homes on concrete blocks.

"Some of those folks can't even pay their electricity bills," Tuck's granny use to chuckle when he was a kid. "But come Christmas, most'a those shacks'll be lit up like Las Vegas."

A large blue heron sailed over the highway in front of the 'Vette. Tuck watched the graceful creature circle a farmer's pond in search of breakfast before touching down on the grassy bank.

"You were born to fly," his granny use to tell him, smoothing a cool hand over his sweaty forehead in that way that told him he was special. "Just like all those aces you read about in your books."

For as long as Tuck could remember, he'd wanted to fly fighter jets. Except for his father, who silently rooted for him, Tuck's grandmother was the only person in his youth to encourage him.

Tuck's mother, on the other hand, use to slip into his room, late at night, when she thought he was sleeping, and whisper in his ear, "Men die in those warplanes, you foolish boy. Get your head out of the clouds and come back to earth."

The 'Vette transported him back to his childhood. Parts of it anyway, to the times spent in Rose Glen, just up the road on the other side of the old iron bridge at Indian Creek. On alternate weekends and every summer vacation of his childhood, his family piled into his Dad's Plymouth and drove the four hours from New Orleans to the backwoods of central Louisiana to visit Grandma Westerfield.

What would Tuck find once he got there? Some answers, he hoped, to a decision—a choice—he would make.

His flight suit drenched, Tuck drove with the window down and his left elbow propped on the ledge. The inside of the car felt like a steam bath. When he bought the 'Vette in '76, he hadn't bothered to get an air conditioner installed because New Mexico's dry climate made the heat tolerable with the windows down. He never imagined he'd live back in Louisiana's sultry climate. He thought he had left for good.

He dropped his speed to forty and frowned when he approached Indian Creek's new, sleek concrete bridge, the old iron one a victim of progress. *So this is what it feels like to get old,* Tuck thought. *When you start to resist change. When you start to resent it.*

Crossing, he glanced into the waters of Indian Creek and his mind tumbled back thirty-eight years.

Barely four feet tall, he and Daddy fished from the bank below the iron bridge where the turtles sunned themselves. He felt a jerk, a tug on the end of his line, more powerful than anything he'd ever experienced.

"Daddy, Daddy, I got a strike," he cried, his scrawny arms pulled back with all their might. "What should I do?"

"Reel it in, Tucker. Reel it in, boy," his daddy hollered and threw down his own rod and hobbled over on his good leg. "You done good, Tucker. You done real good."

Together, Tucker and his daddy pulled the bass from the water, and Tucker held it up like a prize. Up on the bridge, a motorist honked in celebration.

Tuck diverted his mind from the creek and back to the road with the concrete bridge in the rearview mirror. *A man had to be careful*, he thought. The woods closed around him on both sides of the highway. *I can get suckered into the past too easily and believe it was all good.*

A half-mile down the road, tall pines gave way to a cluster of junk shops and small wood-frame houses on the outskirts of town. An eight-foot-tall, plywood alligator held up a sign that read: *Welcome to Rose Glen, Home of the Rose Glen Gators.*

He saw the roadside barbershop where they used to get haircuts boarded up, and the red, white, and blue spiral pole removed from the stoop. Farther into town, a new mini-mart with six Chevron gas pumps replaced the orange-and-blue Gulf station where they used to buy soda pop.

At a flashing yellow traffic signal at the center of town, Tuck turned left onto Main Street. American flags lined both sides of the street for Independence Day. A handful of people milled about, doing Monday morning errands. Tuck cruised past City Hall, a tiny brick building where his grandmother became the town clerk after her husband's grisly death at the sawmill. Tuck's father was seventeen, an only child, and the star quarterback of the Gators football team when the accident happened.

At the old movie house on the corner, Tuck turned right and drove two blocks down a gravel road until he came to the Baptist cemetery. Weathered headstones like broken teeth reached skyward. A statue of Jesus with his palms opened and his arms extended, stood facing Tuck, as if to say, "We've been expecting you."

Tuck got out of the car and went in search of two flat marble markers embedded somewhere in the lawn among the taller, more prestigious pillars.

He found them at last. Almost twenty years had passed since his commanding officer at pilot school in Arizona granted him leave to attend his grandmother's funeral. That was the last time he had set foot in Rose Glen.

Wiping sweat from his brow, he bent down and brushed grass clippings off the inscriptions on the two markers:

Belle Harkins Westerfield
1903-1971
Tucker Foster Westerfield
"Fossy"
1900-1935

"I did it, Grandma," he stared at his grandmother's headstone. "Just like you said I would. But it's not fun anymore."

He sat back on his haunches and gazed up at the sky.

The glint of an airliner flying over at 30,000 feet caught his eye. He squinted, shielding his eyes from the sun, and watched the silver dot until it disappeared.

Lately, Tuck tossed around the idea of flying for the airlines. Jeff Sweeney called him from England six months ago and gave him the scoop on which airlines would hire and how to apply.

"I'm retiring, old boy," Sweenedog said over the phone. "As soon as I get back to the states. It's time I give Jeri and the kids some roots and make some decent money for a change. They don't kiss ass in the airlines, either. Promotions are based on seniority. Not politics."

Tuck studied the date of death on his grandfather's tombstone, and then patted the chest pocket of his flight suit where he kept the 1935 silver dollar. He remembered the shock that summer at the age of seven, when he ventured into the cemetery for the first time and discovered his very own name carved in a place of death. The experience felt creepy.

Confused, he wet his pants and ran all the way back to Grandma's house, certain his grandfather's ghost was right on his heels.

"Somebody put my name on Grandpa's tombstone," he gasped and burst into the kitchen.

Belle Westerfield stood at the stove stirring a pot of gumbo. She chuckled good-naturedly, put the spoon down, and rested her hands on her grandson's shoulders.

"No, Honey Boy. That's not your name there," she explained. "Actually, it is your name, but also your grandpa's name. He just always went by Fossy. That's short for Foster. His real name was Tucker. Tucker Foster. Like you."

"Oh," Tucker sniffled and peeked over his shoulder, for the ghost was out there, beyond the safety of his grandmother's kitchen, to drag him back to the cemetery.

"Naturally, your daddy wanted to name you after your grandfather. Since you were the first born and all." His grandmother eyed him suspiciously. "What were you doing down there anyway? The cemetery's no place for a boy."

"He'll get me," Tucker cried, shaking all over.

"Who'll get you?" Grandma looked past him.

His teeth chattered with fear. "The ghost."

"Oh, Tucker Boy," she laughed and comforted him against her big bosom. "There's no such thing as ghosts."

Tuck filed the memory and left. Minutes later he parked in front of a white clapboard house five blocks away. He stayed in the car, and let the engine idle. Everything looked about the same way it had four days after his grandmother's funeral, when he stood beside his father and watched him hand over the key to a Realtor.

Tuck strained forward, leaned over the steering wheel, and gazed up at the pecan tree next to the porch. A tire swing dangled from the branch where the bird feeder used to hang.

Tuck forgot about the heat, the humidity, the sweat pouring down his neck, as he sat in the car, lost in the past. A yellow Tonka dump truck sat near the edge of the porch at the very spot where he used to jump, his arms extended like wings, into a patch of clover. Nine times out of ten, his grandmother bustled outside in her apron and flapped her arms like a chicken.

"Come on, Tucker," she crowed, and took off down the steps to join him. "Let's you and Granny fly right outta here and chase the clouds away."

Together, the two of them ran around the tiny yard and made airplane noises until the old lady ran out of breath, fell back against the porch steps, and laughed like a schoolgirl.

She was his ally. His confidant. His friend. A pincushion between his mother and the darts of fear she threw at him. Sometimes Daddy, stricken with polio when Tuck was a baby, gimped across the yard to join them. While Mother glared in disapproval over the top of her library book and grumbled, "Shame on you two for encouraging him." And Bo, his mildly retarded brother, with a hearing aid plugged in each ear and thick, black glasses strapped to his head, leaned against the pecan tree and poured over comic books.

Tuck closed his eyes and recalled the hours after his grandmother's funeral, when his mother got smashed on spiked punch and blurted out like a shrew:

"Are you happy now, Belle? Your grandson's gonna go fly those pointy-nose jets and get killed in Vietnam."

That night, Tuck helped his daddy sort through Grandma's things while Bo locked himself in the bathroom with girlie magazines, and Tuck's mother checked herself into the Rose Glen Motor Lodge to sleep off the liquor. They never learned who spiked the punch.

Tuck gunned the engine. He remembered the guilt he felt growing up when he secretly wished he belonged to a different family. A family where the dad had two good legs and could play tackle football and the mother supported her oldest son's dreams. A family where a brother didn't ride a Special Ed bus to school, and a visit to grandmother's house included a grandfather, instead of a cold tombstone a few blocks away. *Such a lousy substitute,* he thought.

After Tuck graduated from pilot school, he vowed he would never return to the state of Louisiana, except to stop through New Orleans once in a while to visit the folks and Bo, who still lived at home. When the Air Force moved Tuck to Beauregard two years back, he had mixed emotions because, thirty minutes away from his front door, the past waited. Until today, he ignored the siren call.

Tuck pushed the gearshift into first and the big engine under the white hood rumbled to life. He spotted the young black woman and the toddler at the screen door, peeping out at him. The same door he traipsed in and out of a thousand times as a kid.

He turned the 'Vette in the direction of home.

Away from a part of his past that nurtured and haunted him. And the dreams he'd finally outgrown. He felt a thousand pounds lighter as he drove back across the bridge.

"This one's for you, Sweenedog." He checked the sky again for the contrail of an airliner that streaked across the big blue ocean of air. A big silver bird taking people places.

"*Come on Tucker,*" Granny urged him on in his head. *"It's time for you to fly outta here and chase the clouds away."*

Chapter Seven

"Schnookums?" Big Sandy parked his briefcase by the front door Monday evening and went to look for his wife.

"We're in here," Wynonna sang out as she spooned chunks of fried hamburger into Baby's dish.

The major ambled into the kitchen. "You run out of Alpo again?"

"No, Daddy. We're celebratin'." Wynonna blew him a kiss, clip-clopped to the sink, and slid the skillet into a pan of sudsy water. "I can't have Baby dinin' on dog food if me and you is feastin' on T-bones."

"What's the occasion?" Big Sandy looked surprised.

"Let's just say my ship came in." Wynonna turned and gave him a frisky grin.

"I take it business was good in Rose Glen?" Big Sandy unclipped his tie.

"Yes-siree-bob." Wynonna bubbled over with pride. "You remember that short, wiry kid who scored all the touchdowns for the Gators last fall?"

"Yeah. Didn't he join the Navy?"

"He sure did," Wynonna giggled. "He's on a big ol' boat

in the Mediter—Meditera—" She gave up trying to pronounce it.

"Mediterranean Sea," Big Sandy interjected and unbuttoned his shirt.

"That's it." She snapped her fingers and smiled. "You remember his mama, Velma? That big ol' gal that runs the motel down there?"

"Sure. She's your biggest fan."

Wynonna clasped her hands. She could hardly contain herself. "Yes, well, Velma's son and a bunch of his shipmates want me to send them the whole kit-and-caboodle in our new line of men's skin care products and colognes."

"The guy sounds like a fag."

"Oh hush." Wynonna slapped him playfully. "You're just jealous."

"Of what?"

"A whole battleship of men begging me for Purple Passion."

Big Sandy rolled his eyes and started to walk away.

"I forgot to tell you, I nearly had a head-on wreck coming home."

"What happened?" Big Sandy whirled around.

"Some lunatic in a blue minivan was drivin' on the wrong side of the road when I came over the bridge this side of Rose Glen. Good thing I was paying attention."

Big Sandy walked back towards her. His tie dangled in his hands. He reached down to give her a hug. "I'm glad you're all right."

"Nearly scared me to death." She tilted her head from side to side to examine his haircut. "Well, lookey there, Clyde made you real purty, but he nicked you somethin' fierce on your ear."

Big Sandy shrugged, looking tired, his mind elsewhere.

"What's wrong, Daddy?" Even in stiletto heels, Wynonna had to stand on her tippy toes to cup his face in her hands. "You don't seem too happy, Angel. I thought my news about Velma's son would cheer you up."

He held her at arm's length and gazed down at her. "I'm real proud of you, Schnookums. It's just that, well..."

"I know," she said in a tender voice. "I've heard the rumors. You thought Westerfield got grounded because you reported him to the wing commander, until you heard he had back problems."

"Another pilot gets away with murder," Big Sandy grumbled and went down the hall.

"Hey, cheer up," she called in her best cheerleader voice. "Colonel high-and-mighty's still grounded, ain't he?"

Wynonna knelt beside the Pekingese and let out a sigh. "Poor Daddy." She stroked the dog's fur. "He should've never joined the Air Force. There's too many dad-gum pilots."

She hunkered down closer to watch Baby eat. "You like that, Angel?" she cooed, taking pleasure in the way his tiny mouth gobbled up the last morsels.

Wynonna pulled herself up off the floor and went into the dining room to clear off the table, laden with hundreds of Purple Passion products. A stack of paperwork sat at one end of the table.

She stashed the paperwork in the bottom drawer of the hutch and put the rest of the stuff on the floor.

Her mind wandered to the day a year ago when she drove home the coveted Purple Passion President's Award. They gave her a customized purple Suburban with her name engraved on the dashboard for being one of the Texas-based company's top-selling consultants.

"Some folks might call Wynonna Sandford a workaholic," declared Ms. Katherine Tuttle, the company's CEO, when she presented Wynonna with the keys at the national convention in Fort Worth. "But Purple Passion ladies know it's a calling—a passion."

Wynonna clung to those words as she set the table with

her best china and crystal. Baby pranced into the room and gave her one of those looks. "You need to go potty, don't you, Angel?" She batted her lashes at him.

Wynonna checked the table one last time. On her way out the door she remembered what Colonel Dennison told Big Sandy that morning.

"I'm sorry, little man, but the wing commander says you gotta be on a leash." She scooped Baby up and went to hunt the leash.

"Daddy." She slammed cabinets and drawers in frustration. "Have you seen that dadgum leash anywhere?"

***

After supper, Gina left Tuck with the kitchen and went to change for the squadron coffee being hosted by the first lady herself, Mrs. Becky Spitz, at her home around the bend on Bayou Way. After touching up what little make-up she wore, Gina pulled a short black-silk sheath from her closet and took it out of the plastic bag.

This'll knock 'em dead, she thought, holding the dress in front of her. *What is tonight's theme? "Summer Casual." A big improvement from the last coffee when the guests were asked to come dressed "cute as a bug." Becky Spitz answered the door that night wearing a ladybug costume, spots and all.*

Gina hated these monthly coffees. They gave her a headache, but she went anyway. Maybe her being there would help Tuck get promoted.

Gina knew she'd get insinuating looks, whispers behind her back around the punch bowl as to why Tuck had been grounded.

She laid the dress on the bed and hunted through the closet for her sandals, still haunted by a dream she had that afternoon when she lay down for a nap.

In the dream, Gina and the boys went with Tuck to a

mortuary to pick out caskets—their caskets—two big ones and two little ones. It was Tuck's idea they would all go together, but he didn't say how. Austin and Jesse stood beside Gina like trusting sheep.

Tuck's daughter stood in a corner laughing at Gina. In his flight suit and cap, Tuck seemed oblivious of the girl. He leaned against a dark mahogany casket trying to convince Gina. "Trust me, Babe, it'll be okay."

Gina's father, wearing a mortician's suit and a big gladhander smile on his face, stood on the other side of the casket. He held a pen and a sheaf of insurance papers.

About to sign their lives away into the hands of her father, Gina jerked the papers out of Tuck's hand, grabbed the boys, and fled.

That's when she woke up.

She bent over to lace up her sandals and tried to analyze the dream.

The dream obviously has something to do with Tuck, she thought. She stepped into a half-slip that barely covered her rear. She kept going back to the part where he said, "Trust me, Babe," the same thing he told her when he left to go see the flight surgeon that afternoon.

The dream is about trust, she decided, *or lack of it. Is Tuck leading us down a primrose path,* she wondered, *or to disaster?*

She shimmied into the black dress, fluffed her hair, and wished she could zap the dream from her memory.

***

Tuck cleared the table, picked scraps of food off the floor. *Now is when we could use a dog*, he thought. The boys played outside in the green turtle sandbox Gina purchased at the BX. He rinsed off a plate, poured the rest of Austin's milk down the drain and stuck the plastic tumbler in the dishwasher while the events of the afternoon unfolded in his mind. *The decision to retire came easy, but facing Gina's bloodshot eyes and cross-examination when I returned from Rose Glen had not.*

Gina hit him with questions the minute he walked in the door: "Why do you have to go see the flight surgeon?—Why is Bull Spitz tagging along?—What's this rumor about you being grounded?"

"Keep your voice down," Tuck whispered, trying to calm her.

Gina leaned against the kitchen counter and hugged herself.

Tuck grabbed her chin to force her to look at him. "You're gonna have to trust me, Babe. The doc's just gonna take a look at my back."

Her eyes narrowed. A tear slipped out. "Your back? What the hell's wrong with your back?"

"Nothing, Babe. Just trust me, okay? That's all I can tell you right now. I gotta go. Don't say anything to anyone."

She dropped her head in her hands. "By the way, Sally called. Michelle flies in tomorrow."

"Tomorrow? I thought she was coming later in the week."

"Me too." Gina grabbed a paper towel to dab her eyes and walked him to the door.

"You look tired," he offered. "Put the boys down and take a nap."

When Tuck got to the hospital, he found Spitz in his Saab, with instructions from Colonel Dennison on what to tell the flight surgeon. Spitz motioned for him to get in the car.

"Your back hurts like hell." Spitz stared straight ahead. "You pulled too many G's over the range this morning. Copy?"

Tuck stared at the man's beaked nose, the crest of snowy white hair. Hunched over the steering wheel, Spitz looked like a big, predatory bird. "So tell me," Tuck said. "Is Captain Goldilocks going to get an early promotion because she rings the boss's chimes?"

Spitz glared at him.

Tuck turned to get out of the car. "Ouch," he howled, grabbing his lower back.

"What's wrong?"

"My back, you asshole," Tuck laughed. "Remember?"

Spitz cursed and got out of the car.

"Where the hell do you think you're going?" Tuck limped across the parking lot.

"You don't think I'm stupid enough to let you go up there alone, do you?" Perching his flight cap on his head, Spitz followed Tuck into the hospital.

"Doc's a good Mormon with six kids," Spitz continued as they headed for the elevator. "You might be tempted to spill your guts."

Tuck swung his head to the side and laughed at Spitz. "You know, Bull, you and Dennison really annoy me."

During the examination, Tuck carried on like a woman in labor. He decided to play along with their charade—for the time being, that is—and wanted to have fun while it lasted. *I'll give Spitz something to squirm about out in the waiting room—something to report back to Dennison that would keep them all guessing.*

"Looks like a pulled muscle," the flight surgeon told Spitz in the waiting room. "I've given him some muscle relaxers. He should lay off flying for a while."

Tuck winced in mock pain and hunched over.

The flight surgeon turned to Tuck. "If it's still bothering you in a few days, I'll need to get some X-rays."

The chime of the doorbell snapped Tuck back to the present. He shut the dishwasher and went to answer the door.

Wheaties bounded into the room, dragging his bashful wife with him. He wore Bermuda shorts and a T-shirt with a picture of an A-10 and the words *Hog Driver* stenciled across the front.

"Howdy, Colonel Westerfield. I came to steal your wife."

Tuck grinned at Sylvia. "Forget about him. You and I should run off together."

"Sir, I don't think your wife would be too keen on the idea." Sylvia Wheaton, with short brown hair and clear, intelligent eyes, looked like the girl next door, neither plain nor pretty.

"Wheaties, you know where the beer is." Tuck turned to Sylvia. "I'll tell Gina you're here."

Wheaties whistled lewdly when Gina strode into the kitchen a few minutes later.

Gina curtseyed. She punched Wheaties' big, freckled arm and reached over to hug Sylvia. "How do you put up with him?"

"It's a tough job." Sylvia shook her head as if wanting sympathy. "But somebody has to keep him in line."

Wheaties grinned at his wife like a little kid. "C'mon, Darlin'." He squeezed her in his massive arms. "You know you're crazy about me."

"Maybe I'm just crazy." Sylvia reached up to give him a kiss.

In the entryway, Tuck nuzzled Gina's ear. "You're lookin' a little too good for a ladies' tea sip, my dear."

"What if somebody says something?" Gina whispered.

"Tell 'em to eat shit and bark at the moon," Tuck whispered back.

On the way out the door, Gina stopped to rattle off a string of instructions.

"The boys need a bath, and make sure they strip outside. I don't want them tracking sand in the house."

"Yes, Boss."

"And don't leave Jesse alone in the bathtub."

"Aye-aye, skipper." Tuck bobbed his head. "I'll make sure he doesn't drown while you're gone."

Gina frowned at his last remark. "One more thing. They need to brush their teeth, but you'll have to help Jesse."

Tuck turned to Wheaties. "See what you have to look forward to when you're an old fart like me." He blew Gina a kiss. "Don't worry, Babe, the boys'll be fine."

"Yeah," Wheaties laughed and gave the ladies a thumbs up. "I'll personally see to it they don't guzzle all the beer or fondle the dancing girls."

Wheaties wandered into the living room and stood before a wall of shelves crammed full of mementos of Tuck's flying career. The collection encompassed everything from model airplanes and service medals to squadron mugs and bronze figurines that depicted military aviators of various eras.

"Most wives wouldn't let their husbands keep all this stuff in here," Wheaties said.

"Gina's not most wives." Tuck cracked a smile.

"I'm glad Gina went with Sylvia tonight." Wheaties took a drink of his beer. "She gets herself all worked up over these silly wives' functions." He set his beer down and picked up a model of an F-4 to examine the plane in detail. "To tell you the truth, Sir, she hates going. She says it's like walking into a hen house."

"Can you blame her?" Tuck said. "The only reason Gina goes is to support my career. She thinks it'll help me get promoted. Just don't tell her it's called brown nosing."

Wheaties laughed.

"Besides, Gina needed an excuse to get out of the house tonight. She had a rough day." Tuck scratched his chin.

Wheaties put the plane back on the shelf. "Because you were grounded, Sir?"

Tuck picked up a model of an F-111, squinted at it, then wiped the dust off the tail section and moveable wings with the bottom of his T-shirt. "Partly, I guess." He pulled the wings back and wiped the tiny windshield of the two-man cockpit where a pilot and navigator sat side by side. "The Aardvark—Cannon Air Force Base." He admired the plane at arm's length. "That's what I flew after the war."

Wheaties nodded.

"You know what we called the Aardvark?" Tuck grinned.

"I just know it's a big-ass airplane." Wheaties shook his head.

"The side-by-side, super-sonic swing-wing voting booth," Tuck joked. "Only plane in the U.S. arsenal where the entire capsule ejects, taking you and your navigator with it."

Wheaties stuck the F-111 back on the shelf. "Just like the Gemini astronauts." He took a swig of beer. "Did you see the blurb in *Air Force Times* about your buddy's crash?"

"Sounds to me like he flew into the ground."

"Sir, I've gotta ask you something." He rubbed the red stubble on top of his head. "I know it's none of my business, but, well, the rumor mill at the squadron worked overtime today."

Tuck picked up one of his air medals encased in a gold frame and studied it. "That so," he said matter-of-factly and turned it over in his hand.

Wheaties tossed his head back and guzzled the last drops of beer. "Colonel Spitz called us all into the briefing room and said you pulled your back out over the range this morning."

"He did, huh?" Tuck raised his brow.

Wheaties picked up one of Tuck's logbooks and thumbed through it. "After the briefing, Spitz left for a meeting. Everybody asked Tony what happened, but Tony wouldn't say a word."

"This has nothing to do with Tony."

Wheaties whistled and held up Tuck's logbook. "This is impressive, Sir. All these hours you've logged." He closed the book and placed it back on the shelf with the same reverence people give to Bibles.

"Tony cornered me in the john after the briefing. He told

me what happened over the range, or didn't happen." Wheaties collected his thoughts. "Tony said your back seemed fine this morning—before and after you landed."

"Wheaties, have you ever noticed how ceiling texture looks a lot like cottage cheese?" Tuck's voice sounded far away as he contemplated the little curds of plaster, his neck bent back as far as it could go.

Wheaties craned his neck. "Well, actually, Sir, can't say as I've ever spent much time thinking about it, but now that you mention it."

"My grandmother always served it with dry toast, mashed potatoes, and sliced tomatoes. That's good eatin'."

"Sir, are you all right?"

"Sure," Tuck shrugged. "Life's a bitch, and then you die."

Wheaties scratched his head. "So how long are you grounded?"

"Till the doc tells me I'm not."

Wheaties squeezed his empty beer can in his huge fist. The crinkle of aluminum filled the room and broke the silence between them.

"Shoot, I hope this isn't because of me." Wheaties punched the air with his fist like a heavyweight boxer. "I know you went out on a limb for me."

Tuck blinked, surprised that Wheaties would think such a thing. "This has nothing to do with you."

"If you say so." Wheaties dropped the subject, walked over to the patio door and gazed out at the boys.

Tuck stuck his hands in his pockets and paced up and down the living room.

"I'm thinking about putting in my papers. I hear the majors still hire old farts."

Wheaties' head snapped around. "Retire? You're kidding me."

Tuck paused in front of a large black-and-white photo of

himself standing next to a F-4 Phantom loaded with live bombs on the ramp at Korat Royal Thai Air Force Base in 1972. "I sure had more hair back then."

"You don't need hair to fly an airplane."

"I'm gettin' too old for this shit. Flying fighters is a young man's game."

"Shoot, Sir. You'll get your own squadron soon."

"Let's face it. A squadron commander's gotta be a politician, and I'm not one. Never have been, never will be. Besides," he shrugged, "if I stick around, I'll end up like those has-beens you see hanging around the club—either because they can't let go or they're scared to try something else."

Wheaties cracked a knuckle. "What does Gina think?"

"I haven't told her yet. Ready for another beer?"

Wheaties followed Tuck into the kitchen.

Wheaties cleared his throat and squared his shoulders. "Don't put your trust in earthly princes," he said, "in mortal men who can't save you. When their spirits depart, they return to the dust, and all their plans perish."

Tuck gawked, dumbfounded. "What the hell is that suppose to mean?"

Wheaties pulled the tab off his beer and tossed it in the trash. "Translated into fighter pilot, it means check six, 'cause you can't trust the assholes around here." He headed outside to see the boys.

One minute Wheaties was a hell-raising warrior, the next a poet.

"Old Testament," Wheaties threw over his shoulder before he stepped out onto the patio and called to the boys: "Hey, little drivers. It's Uncle Wheaties."

***

At the Bolton Mall earlier that evening, clad in scruffy jeans, a worn-out golf shirt, cheap sunglasses, and an old baseball cap that belonged to his son, Paul, Colonel Glen Dennison locked his wife's Volvo and strolled into Sears. A few minutes later he came out of J.C. Penney's on the opposite end of the mall.

He went to Linda Garrett's light-blue minivan parked nearby. He unlocked the door using the key she slipped him at work and drove around to Dillard's.

At the curb, she opened the door and crawled in. She wore a plaid cotton shirt, khaki shorts, and tan flats that enhanced her thin, athletic calves. With her hair cut in a pixy and an impish face dusted lightly with powder, Linda Garrett looked like a high school gymnast, not a woman in her thirties. She dumped her bags in the back seat and they drove out of town, heading south. The sun wouldn't set for another two hours.

"Where did you tell Lois you were going?" She adjusted her shoulder strap.

"The bookstore. She's going to a coffee tonight, the 428th's. What about Hank?"

"He took the boys to baseball. They're both on the same All-Star team." She turned sideways in her seat. "I need to be home by nine-fifteen. The mall closes at nine. Hank will get worried if I'm late."

"Relax." Dennison's hand slipped off the steering wheel and up one leg of her shorts.

She yawned and rested her head against the seat. "Where are we going?"

"A motel not too far from here. I spotted it while Lois and I were out for a Sunday drive. Now get some rest. You look tired."

Dennison drove one-handed down the skinny, two-lane highway into a thicket of piney woods. Thirty minutes later, a bridge loomed ahead.

"Wake up." Dennison nudged her. "We're almost there."

She yawned, unfazed by his roaming fingers. "How long was I out?"

Dennison looked up just in time.

"Oh shit," he yelled. "Duck."

A purple Chevrolet Suburban barreled toward them on the narrow bridge and headed straight for the minivan.

Dennison swerved back into his own lane just as the Suburban zoomed past and blasted its horn.

Linda Garrett twisted around in her seat. "Wasn't that Major Sandford's wife?"

Dennison glanced nervously in his rearview mirror. "What the hell is that bubble-headed blonde doing clear out here in the middle of bum-fuck Louisiana?"

Dennison slowed when they passed a plywood alligator by the side of the road on the outskirts of town. "It's a quaint little place," he said.

Moments later the minivan came to rest in front of the Rose Glen Motor Lodge, a seedy, L-shaped, single-story structure.

"This place looks like a flea bag." The captain wrinkled her nose.

Dennison ignored her remark. "You stay here. I'll get the key."

A bell jingled when he stepped inside the motel office and closed the door. A hefty woman with dark circles under her eyes reading a travel brochure sat behind the front desk. An air conditioner blew frozen air out of a window unit.

"Evening, I'm Velma." The woman put down the brochure and got up. She had on a pair of old trousers and a man's work shirt. A romance novel lay opened face down on the desk, the book's spine cracked under the strain.

Dennison glanced at the brochure, then flashed her a big country grin. "Taking a trip to the Middle East?"

Velma didn't crack a smile. "My boy's over there. He's in the Navy. You alone?"

Dennison chuckled easily, running a hand through his hair. "I'm with my wife. We'd like a room for the night." He extracted a wad of cash from his wallet and noticed a stack of purple-colored business cards on the desk. His eyes lingered there a little too long.

"Take one for your wife." Velma pushed a card in front of him. "That Purple Passion puts the zing back in a woman's thing." She pointed a short, stubby finger at the name on the card: Mrs. Wynonna Sandford, Beauty Specialist.

Dennison picked up the card and tried to steady his hands. He felt the blood drain from his head then stuck the card in his wallet and forked over the cash.

"You're probably wondering why I don't wear the stuff myself," Velma chuckled, ringing up the ancient cash register. "Let me put it this way: My zing hit the road a long time ago. And it ain't comin' back. But let me tell you," she tapped her finger on the counter to make a point, "it's shore worked wonders for the other ladies in town." She winked knowingly and handed him the key. "Did I mention you help our football team out, too? Mrs. Sandford donates a portion of her profits to our Booster Club. She's a real fine gal. Her husband's in the service, you know."

Dennison shook his head and dropped the key in his pocket. His mouth went dry.

"Now you give that card to your wife," Velma instructed. "Tell her to help herself to the samples on the night stand."

"I certainly will. Thank you, Velma." He opened the door to leave.

"Oh, I didn't catch your name," she called.

Hesitating, Dennison touched the bill of his cap. "It's Paul, Paul Davis." He started to close the door.

"Mr. Davis?"

"Yes?" He stuck his head back inside.

"Wynonna carries a wonderful line of men's products as well. You give her a call. Tell her Velma sent you." She picked up the travel brochure and buried her face in it.

Five minutes later, Colonel Dennison and Captain Garrett

rolled around on a lumpy mattress in a dark, musty room.

"Be good to me, Linda," Dennison breathed, "And you'll get that early promotion to major."

On their way back to Bolton, Captain Garrett opened a small tube of hand cream and squeezed a drop in her hand.

"That was a close call on the bridge earlier." Dennison watched her work the lotion in. "We better refrain from doing this for a while. Marsh was all over my ass this morning."

"You don't think Major Sandford's wife saw us, do you?" She drew her hand to her nose. "This smells nice. I may have to order some. Wanna smell?"

Dennison cleared his throat. "Actually, Lois keeps a tube of that stuff by the kitchen sink."

Linda Garrett's face fell. "Oh." She rolled down the window and threw the lotion out.

"What did you do that for?" Dennison frowned.

She shrugged. "Does Lois suspect anything?"

"Are you kidding?" Dennison chuckled. "She's too busy playing Mrs. Wing Commander to know I exist."

"She'll make a good general's wife. She would die, you know, if she knew who protected you."

"Don't worry about that. Marsh is the last person on earth who would tell her."

"Do you think Westerfield will keep quiet?"

"As long as I keep promising him that squadron—against my better judgment."

"All the young guys like him," the captain said. "Especially Hank."

"I know that. That's precisely Tuck's problem."

"What do you mean?"

"He's too popular," Dennison stated. "You cannot be a leader of men and expect to remain their friends. On top of that, he's too much of an idealist." The word rolled off his tongue with repugnance.

Dennison fumbled for the headlights on the unfamiliar

dashboard. "I'm a bit concerned though. Spitz said Westerfield acted rather peculiar at the flight surgeon's today."

"Like how?" She pointed to a silver knob next to the radio.

"Like it was all a game." Dennison flipped on the headlights and settled back in his seat.

"Isn't it?" Captain Garrett remarked quietly while the trees turned to dark shadows in the night.

Chapter Eight

On a muggy Tuesday afternoon, the third of July, a commuter plane taxied up to Gate One, a simple gate in a chain-link fence at the Bolton Municipal Airport. A drabby, outdated terminal stood nearby.

Tuck curled his fingers around the fence; trickles of sweat rolled down his temples.

"When's the last time we saw her, Gina?" He breathed in the thick, stagnant air.

Gina dabbed her forehead with a tissue. "We were still in Alaska. Austin doesn't remember her, and Jesse wasn't born yet. About four years, I guess."

Tuck looked down at the boys, at their ruddy faces pressed against the fence. Gina had them spruced up in crisp summer short sets and neatly combed hair.

Hope they all get along, Tuck thought.

After the pilots shut down the engines, the hatch opened and a flight attendant released the folding stairs out the side of the plane. Two men in business suits and an old lady with a knitting bag got off first.

Tuck's anxiety mounted each time another passenger filed out. A young mother with a baby. A two-striper from the air base. A tall, willowy teenager with vampire make-up and punk hair, lugged a boom box and a duffel bag slung over one shoulder.

Tuck stiffened in shock.

Gone were the silky blonde locks he remembered, the fly-away tresses that seemed to have a life of their own. He stared at the kid's butchered hair, streaked black up the sides. Michelle had on men's underwear—paisley-printed boxers worn over Spandex bike shorts, a ratty white T-shirt, and combat boots.

Tuck gulped, butted his head against the fence. "She looks like a cross between Alice Cooper and Madonna."

The girl hesitated, took a deep breath then clomped down the stairs and plodded across the hot black tarmac. The closer she came, the tighter Tuck clung to the fence. Michelle's flimsy white T-shirt left little to the imagination. She was big busted like her mother and just as carefree and uninhibited about it.

Tuck backed away from the fence. "Jesus," he groaned. "She's not wearing a bra."

He shut his eyes, hoping she was an illusion. But when he opened them, she was as real as the amused looks on her fellow passengers' faces and the angry music that screamed out of the portable stereo at her side.

"She looks ridiculous," Gina said.

The girl drew near and wide-eyed Austin looked up at Tuck. "Dad, you think she has fangs?"

The girl reached down and turned off her stereo. Her lips, the color of old blood, curled in a sneer.

Jesse giggled and clapped his pudgy hands. "Yippee. It's Halloween."

Tuck scowled at Gina over the rim of his sunglasses. "She's chopped off her hair and dyed it like a skunk. And those boots, I can't believe Sally let her out of the house dressed that way."

Gina shot him a disgusted look. "What were you expecting, Tuck, the Virgin Mary?"

Tuck felt wounded. "That's not fair, Babe. Just try to be gracious, okay?"

"Whatever you say," Gina sighed. "Just remember. I'm no Mother Teresa."

He took Gina's hand. "C'mon. Let's kill her with kindness."

"Anybody got a clove of garlic and a crucifix," Gina muttered, dragged her heels when they stepped forward to greet the girl.

Jesse scrambled out in front of them and giggled with delight.

"Trick-or-treat. Smell my feet," he blurted and flung himself at the stranger.

***

The next day, they loaded up the Bronco to head into town. Tony and Krystal Grimes were having a pool party at their new house in Bolton, and the boys could hardly contain themselves.

"All aboard," Tuck called, throwing in the lawn chairs.

In flip-flops and swimming trunks, Austin and Jesse clambered out of the house and piled into the truck. "I'm gonna jump off the diving board." Austin buckled himself in behind the driver's seat.

Gina reached in to strap Jesse into his car seat. "What if you forget how to swim?"

"Then I'll dog paddle," Austin said, as if this was the most obvious thing to do.

"But what if you panic?"

"If he panics," Tuck said through the rear window, "all he has to do is float like a turd in a punch bowl."

"Turd in a punchbowl. Turd in a punchbowl," Jesse chattered, waving a Ninja Turtle in Gina's face.

Tuck finished stowing their gear. "Where's Michelle?"

Gina signaled toward the house. "She's sulking because I told her to leave her boom box here. Along with that heavy-metal crap she listens to."

Tuck shut the tailgate and came around to the driver's side. "Don't tell me you're not a big fan of Guns N' Roses?"

Gina eyed him over her sunglasses. "Let's just say Tony and Krystal might find some of their music offensive. If you get my drift."

Tuck nodded. Gina referred to the n-word—a word tossed out at random during Tuck's childhood, mostly by southern white folks who didn't know any better, or too crass to care.

"So much for Sally's Jesus music," Tuck snorted.

"While St. Sally's away, the devil will play," Gina mumbled when Michelle came out of the house.

She wore the same outfit from the day before, including the ghastly make-up. She shambled along in her battered boots toward the Bronco, where Gina stood by the passenger door, waiting for her to get in. Suddenly, Michelle's left foot stepped off the pavement, crushing the cluster of marigolds growing by the driveway.

Gina gasped when Michelle's inky lips curled in a sneer.

"Sorry." The girl pretended to survey the damage. She lifted her foot out of the flower bed. "Guess I lost my balance." Smirking, she climbed into the Bronco and tracked dirt on the floor mats.

Gina seethed at the flash of scorn in the girl's green eyes, at the stubborn cleft chin, squared off in defiance. In Gina's mind the act was an attempt to get back at her for prohibiting the girl from taking the boom box to the party.

The potent smell of mashed marigolds drifted up to Gina's face and riled her even more. "Try to be gracious," Tuck suggested.. "Let's kill her with kindness."

Kindness my foot, Gina thought and eyed the giant bootprint in the orange and yellow flowers. Gina climbed into the Bronco and slammed the door. What the girl needed was a swift kick in the rear.

***

Tuck fired up the engine.

"Who's ready to party?" He glanced at Gina. She sat tight-jawed in her tank top and running shorts and clutched her fist as if she wanted to punch somebody.

Stumped, Tuck glanced in the rearview mirror. Sandwiched between Austin and Jesse, Michelle simpered over something. When she caught Tuck watching her in the mirror, she averted her eyes. Even under all that goo, Tuck could see her blush.

He took a deep breath and cranked up the air conditioner. He didn't want any confrontations. Not this early into her visit.

He backed the Bronco out the driveway, and they headed down the street. They rumbled along in awkward silence over the blackened waters of the bayou, where gnarled cypress trees bent over the edge of the banks like women doing wash. Above the murky surface, hundreds of dragonflies hovered.

They motored out the back gate, past the gate house in the middle of the bridge, where an armed guard conducted ID checks on all the cars coming over the bayou onto base.

Picking up speed, they zoomed down a narrow stretch of blacktop that sliced through a flat, spongy section of farmland. In the fall, a pumpkin patch sprang up along the roadway and created a blaze of color. A huge flock of noisy crows took over the fields, cawed and pecked and turned the countryside black. In the distance, the woods huddled next to the fields like a crowd gathered at the scene of a crime.

While Tuck glanced at the speedometer, a loud k-thunk crashed down on the center console between the front bucket seats. He jumped, jerked his head around. "What the hell was that?"

Michelle propped her feet on the console and crossed them when he turned back and fixed his eyes on the road.

"There's not a lot of leg room back here," she grumbled. "Unless you're a midget."

She wiggled her foot and the console rattled like an old tackle box. Tuck grimaced. *Either the kid is nervous, agitated or has to pee.*

Beyond the windshield, beyond the white stripes flashing by in front of him, a red dot appeared on the horizon, and Tuck slowed as they approached the stop sign. Gina clutched the door like she was ready to bail.

After checking both ways, Tuck pulled onto a busy four-lane highway and headed into town. Along the way, they passed a string of used car lots, a handful of pawn shops, a junk yard, and an old-fashioned burger joint painted red, white, and blue. A sign out front said: "Ten percent off with military ID."

He glanced sideways. "You okay, Babe?"

Gina shrugged and stared out the window, her shoulders humped like a fortress.

The console rattled on.

Tuck figured he better say something to Michelle, quick before Gina blew up. When Gina's nerves were frayed, little things set her off.

Tuck gripped the steering wheel. The last thing he wanted to do was hurt the kid's feelings. He threw a quick glance over his shoulder.

"Honey, sweetheart," he lavished her with endearments, "that gets annoying while I'm trying to drive." He dropped his gaze to her boots.

Michelle picked her feet up and slammed them down on the floorboard. "What is this? A goddamn prison."

Sally will have a fit, Tuck thought as a carload of airmen blew past him in a metallic blue Trans-Am.

And Jesse would have another cuss word to add to his repertoire.

Before Tuck could think of a comeback, Gina twisted around in her seat. "No, Michelle. This isn't a prison. We're all one big happy family. And since today is a holiday, we're going to have a good time. Even if it kills us." She turned around, crossed her arms, and looked smug.

Tuck glanced in the rearview mirror, then fixed his eyes on the traffic signals, the side streets into various neighborhoods, the elderly couple out for a stroll.

Anything to take his mind off the pitiful image of his daughter; her fingers looped around each other as if in prayer. Her long, elegant legs scrunched up and the toes of her boots turned inward. Her downcast eyes gave nothing away.

Tuck swallowed, flushed with shame, at the sudden realization that for a tall girl like Michelle, a back seat couldn't accommodate her long limbs, unless she rode around in a limo.

He decided that on the way back, Michelle could put her feet any place she pleased. Until then, they had to suffer. Now wasn't the time to backpedal and apologize.

Again, an awkward silence swept over the interior of the truck. The only sounds were the tires rubbing against the highway and the air conditioner blasting out cold air.

After a while, Austin hummed and banged his head against the back of the seat.

Tuck had a good view of Austin in the rearview mirror and cleared his throat. "What'cha humming, son?"

Austin froze. "The ants song," he answered meekly.

"Which one are you on?"

"Number five."

"Mind if I join in?"

Austin shook his head and thumped it again. "You go first, Dad."

They sped through another green light, and Tuck opened

his mouth to sing. "The ants go marching five by five, Hoorah, Hoorah..."

To Tuck's relief, the whole family joined in.

A short time later they rumbled into Good Earth, an upper-middle-class neighborhood on the west side of Bolton. Half the homeowners were military, mostly midlevel officers and senior NCOs who opted to plunk down a wad of cash on a down payment and a monthly mortgage, in exchange for a cost-of-living-allowance and a nicer home than the Air Force provided on base.

The Bronco rolled down Tara Lane and slowed in front of a rusty-brick ranch where a bundle of balloons waved from the curbside mailbox. Jo-Ellen's New Yorker and the Wheaton's dusty Jeep were parked at the curb.

"Looks like we're the last ones here," Tuck said over a cacophony of voices from the back seat.

Gina chuckled when she saw the two compact cars parked inside the garage. "They'll have to trade in one of those toys if they ever have kids." Tony's white BMW sat next to Krystal's red Mazda RX-7.

Tuck slid the gear shift into park and pocketed the keys in his khaki shorts. "It'll never happen. Krystal's too much of a career woman."

Tuck watched Gina slide out of the truck. She raked a hand through her hair, admired the house and neighborhood. She twirled around. "One of these days I'll have a real house," she said. "A house with a big front porch, hardwood floors and a big stone fireplace where the kids can hang their Christmas stockings." Her eyes danced with promise.

"Hurry up," Jesse wailed.

"It's hot back here," Austin grumbled.

Tuck closed the door and walked to the back of the Bronco. "Gina, you're the one who wanted to live on base." He swung a lawn chair through the air. "Because you're closer to the commissary and the emergency room."

"I know, it's just that, well, as many times as we've moved, if you've seen one set of quarters you've seen 'em all. The same boxy rooms, the same ugly linoleum." She wrinkled her nose. "I swear, Tuck, the same cockroaches follow us from base to base."

"Mommy," Jesse screeched and struggled to free himself from his car seat.

"Screw this." Michelle threw down her seatbelt, unbuckled Jesse, and handed him to Gina.

Tuck carried a pan of baked beans around the side of the truck. "When I retire, we'll build you that dream house. Be careful, they're hot."

Gina set Jesse down and took the beans. "But that's going to be awhile. After you make full colonel."

I'm not sticking around that long, Tuck wanted to tell her.

Michelle got out of the Bronco. "Bitch, bitch, bitch." She bent over to brush dirt off her boots. "At least you live in a house." She straightened back up.

Gina flinched, gripped the pan of beans.

Michelle grabbed Jesse's hand and started to take off. "Cockroaches are nothing compared to the cocksuckers that live in my apartment complex."

"Count to ten," Tuck told Gina then hefted a lawn chair under each arm.

Plunking the beans down on the broiling hood, Gina hissed, "Son-of-a-bitch," and kicked the tire.

Austin slid out of the truck. "What's a cocksucker?"

"Never mind," Tuck followed him up the driveway.

An attractive black woman in a green bathing suit and a chiffon cover-up answered the front door.

"Colonel Westerfield, where are your swimming trunks?" exclaimed Krystal Grimes and ushered them inside.

Tuck stooped to give her a hug. "Krystal, the only part of my anatomy that's getting wet are my lips on a cold beer."

"Well, Happy Fourth of July to you, too," Krystal shot back, her handsome mahogany face split into a broad grin. Krystal worked at a law firm in Bolton as the legal assistant to one of the senior partners. Her dream was to attend law school when Tony got out of the Air Force.

She turned and winked at the boys. "I've got a surprise for your crotchety old papa."

Austin giggled. "Are you gonna throw him in the pool, Miss Krystal?"

"Bingo." She threw her head back in a laugh.

"Bingo," Jesse mimicked and clapped his hands.

Krystal tickled Jesse's tummy. "If the spirit won't move him, we'll have'ta baptize him ourselves."

Michelle stood to the side, looking bored. "Good thing my mother isn't here. She'd accuse you of blasphemy."

Krystal turned, slightly amused. "You must be Michelle." She extended her hand. "Glad you could join us, dear."

Michelle's dark lips twitched nervously, and she shuffled forward for Krystal's hand. "What I meant was, my mother's a holy-roller. And, well…"

Tuck cleared his throat. Michelle and Krystal looked up expectantly. "Shall we join the others?" he suggested and picked up the lawn chairs. Talking about Sally made him uncomfortable, especially in front of the boys.

Michelle gave him a loopy grin, then turned to address Krystal. "You know what my mother says about him?" She flicked her head at Tuck. "She says he's possessed."

Krystal let out a chuckle and winked at Tuck.

He rolled his eyes and shook his head.

"Krystal, where do you want the baked beans?" asked a frazzled Gina from the rear.

"Furniture comes later." Krystal shepherded them through a semi-empty den. "When Tony gets that pilot bonus for signing on with the Air Force another four years."

She threw back the sliding glass door onto a large,

covered patio. A ceiling fan whirled above a huge picnic table laden with food. A bug zapper hung nearby. The sweet, soulful tenor of Aaron Neville crooned from two gigantic speakers hauled out from the den.

Tuck's mouth watered at the tangy aroma of grilled chicken, a good sign his appetite returned. Since his involvement in the cover-up, he dropped five pounds.

Tony looked up from the grill and waved a spatula in their direction. He did a double-take when he saw Michelle.

"It's about time you characters showed up." Grinning, he pointed the spatula at a couple of ice chests on the patio. "Help yourselves. There's beer, wine coolers. Soda pop for the kids."

Under the shade of the diving board, Wheaties and his wife pecked away like two lovebirds. Without looking up, Wheaties stuck a hand in the air and waved.

***

While Tuck beelined it to the cooler for a beer, Austin walked to the edge of the patio and gazed at the swimming pool.

Gina peeled the top off a wine cooler and took a sip when Jesse trotted up in his swimming trunks. He thrust his scrawny chest out. "Take off your clothes, Mommy. I want to go swimming."

"In a minute," Gina replied, working her way to Jo-Ellen and her short, stocky husband, Buzz. With his '50s burr cut and build, Chief Master Sergeant Buzz Hawkins reminded Gina of Sergeant Carter from the old *Gomer Pyle* show, except that Buzz didn't have Sergeant Carter's temper.

The Chief tossed down a handful of peanuts when he flagged her over. "Hey, good-looking."

Jesse went berserk. "Take me swimming," he tackled Gina from behind.

"Knock it off, Jesse. Mommy wants to say hi to these folks." Gina reached down with her free hand to untangle herself. Jesse refused to budge and clutched her calves even tighter.

From across the patio, Austin yelled, "Shut up, Jesse. Mom's already in a bad mood."

Popping the top off a can of grape soda, Michelle came to Gina's rescue. "Hey, settle down, Squirt. Your Mom said she'd take you in a minute."

Like an angry young bull, Jesse dropped his hold on Gina and charged after Michelle. He rammed into her knees. "Don't call me Squirt."

The can of grape soda got knocked out of Michelle's hand, plopped to the ground like a purple fizz bomb, and gurgled its contents all over the patio.

"Jesse, look what you've done," Gina snapped, appalled at her son's behavior.

Jo-Ellen grabbed a roll of paper towels and Buzz ran to get a pitcher of hot water.

Austin clutched his beach towel to his bare chest. "Jesse, you're such a brat." He turned his back on his brother.

Michelle licked grape soda off her hand and squatted down in front of Jesse. "Tell you what. If you apologize to your mom, I'll take you and your brother swimming."

"Jesse can't swim," Gina blurted, uneasy by the girl's sudden offer to help.

Michelle blinked and lowered her gaze in expected defeat.

Michelle's expression caught Gina by surprise, and she gave in. "You'll have to watch Jesse like a hawk. He's not allowed in the water without his arm floaties."

Buzz gave Jo-Ellen a hand when Tuck walked up. "What happened?"

"You're a day late and a dollar short." Jo-Ellen rushed off with a dripping wad of paper towels for the nearest trash bin.

Tuck stepped back as Buzz moved in to douse the area

with water before it got too sticky.

Gina barely acknowledged Tuck; her focus was on Michelle. "Austin just learned how to swim. Don't turn your back on him for a second."

"But didn't Mom tell you?" Michelle unzipped her fanny pack and whipped out a certificate from the Red Cross. "I'm a certified lifeguard. I passed my driver's ed class, too." She showed them her Texas driver's license.

A wave of sadness rushed over Gina. She glanced at Tuck.

He stroked his chin, suddenly looking old and distraught. "Why didn't your mother tell me you drive? Or that you had a job?"

Michelle zipped up her fanny pack. "You didn't ask." She kicked at the concrete with the toe of her boot. "She made me quit my job after the first week anyway."

"Why?" Gina asked.

Michelle's head snapped up. "For wearing a bikini to work."

"What's wrong with wearing a bikini?" Gina peeked down her own shirt at the top half of her bathing suit.

"Bikinis are for hussies, Mom says." Michelle smirked. "Not women of virtue." She took Jesse's hand and left.

"Oh Jesus," Tuck growled, rolling his eyes at the others.

"Well, heavens-to-Betsy," Gina crowed then slapped herself on the thigh. "Just call me the Whore of Babylon."

"Poor kid," Jo-Ellen remarked after Tuck ambled off to set up the volleyball net. "I feel sorry for the girl. What kind of mother would say those horrible things to a daughter?"

Buzz dug into a bowl of potato chips. "A crazy one, that's what." He crunched on a chip. "Or a woman who's jealous of her own daughter."

Gina kept an eye on the yard at the shallow end of the twelve-foot diving pool, where Jesse stood at the edge in his orange arm floaties, ready to jump in. Austin swam circles around Michelle.

The girl's boots, socks, and fanny pack were tossed aside in the grass next to a lawn chair.

"Looks like Michelle's got everything under control," Jo-Ellen remarked cheerfully.

Gina sipped her wine cooler. "Jesse thought she wore a Halloween costume when she got off the plane."

Jo-Ellen puffed on a cigarette. "I see kids like Michelle all the time at school. In clothes even the Salvation Army would reject. She's just going through a stage. It could be worse."

Gina ran her finger around the rim of her wine cooler. "What could be worse than all that crap she's got on?"

"She could be pregnant." Jo-Ellen blew smoke out of her mouth.

"Or on drugs," added Buzz.

"Look, Honey," Jo-Ellen scissored her cigarette then pointed it at Gina, "Just because she looks like death warmed over doesn't mean she's a bad kid. Look at the way she's got your boys wrapped around her finger. They mind her, too."

Gina turned to check on the kids. "I guess you're both right," she sighed, before her eyes zoomed in on the girl's wet T-shirt plastered against her enormous breasts like a second layer of skin. Suddenly, Gina felt like an old prude. "I certainly hope Michelle threw in a change of clothes. Her tits are showing."

"Look at it this way," said Buzz, being diplomatic about the whole thing. "At least she's keeping her head above water. Can you imagine how she would look if her make-up started to run?"

Jo-Ellen poked him in the ribs. "Behave yourself, Buzz. We both know you're not looking at her face."

Buzz Hawkins slung his arm affectionately around his wife. "Now whatever gave you that idea?"

***

Tuck wrestled with a pole to the volleyball net. Each pole came in two pieces for easier storage. Tuck couldn't get the two ends to join up.

"Let me give you a hand with that." Tony bent to grab the other end of the pole. Together they popped the two pieces into place. "Don't want you to strain your back, Colonel."

Tuck eyeballed his friend. "Cut the bullshit, Tony."

"Just what I thought," Tony shook his head at the ground. "Then why did Colonel Spitz give us that song and dance about your back?"

Tuck rummaged through the box and pulled out the other two fittings and slid them into place. He ignored Tony's question.

"Well, I guess it's none of my business. It's just that, well, Wheaties and I are a little worried about you, that's all."

Tuck went to unravel the net.

Tony hesitated, then got up. "I better go turn the chicken. Sure you're all right?" Stuffing his hands into the pockets of his Bermuda shorts, the captain rocked back and forth on the heels of his sandals.

Tuck peered up at his friend's worried face. "Trust me, Tony. I'm fine, really."

"Sure you are," Tony walked off. "And I'm a white boy, too."

Tuck walked back to get a fresh beer. Wheaties climbed out of the pool and loped over to join him.

"You got a great kid there, Colonel Westerfield." Wheaties dried his face on a towel.

Tuck glanced at the pool. "Oh, you mean Michelle."

Wheaties draped the towel around his thick neck and followed Tuck's gaze. "She's just working through some stuff right now is my guess."

Tuck turned and noticed the lieutenant's pale skin, pink now from the sun. The scent of chlorine lingered in the air, reminding Tuck that he had once been a swimmer, flying through the water on his high school swim team, before he abandoned the butterfly and learned to fly through the air. "Did she tell you that?" Tuck glanced back at the pool.

Wheaties grabbed both ends of his towel and snapped it over his head. "She didn't have to. It's written all over her face. Heck, Sir. When I was her age, I grew my hair long, stopped going to church, and even dropped out of football for a year."

Tuck stared back at him, surprised. With his muscular body, square face, and red bristles, the lieutenant could pass for a Marine drill sergeant. "So what's with the war paint?" Tuck asked, turning to look at his daughter as she gave Jesse a horsy ride across the pool.

By now, Michelle's make-up melted into an oozing mask of black and white grease.

"Teenage rebellion. Pure and simple," Wheaties said.

"Uncle Wheaties," Jesse screeched over the top of Michelle's head.

"Get back in the pool." Austin joined in, riding around on Sylvia's back.

Through a spray of water, Michelle yelled at Tuck, "Hey, old man, afraid you'll mess up your hair?"

"What hair?" Tuck laughed to cover the ache in his throat.

"Hey, Driver, watch this," Wheaties winked, bounded off to ham it up for the boys, and left Tuck alone to drown in the wake of his young friend's energy.

In a daze, Tuck stared at his sixteen-year-old daughter, embarrassed that he could see her breasts through her wet T-shirt. He had a sudden urge to throw a towel over her chest.

His throat closed which made it difficult to swallow. *Yeah, the child I failed, the child I almost lost.*

Overcome by guilt, Tuck averted his eyes and recalled the day Sally popped up at the squadron, months after the divorce, pregnant as an elephant. When he asked if the kid was his, she belted him with her purse and took him to court for child support.

Once the first check cleared the bank, she told him to keep away from their daughter or God would strike him dead.

Years later, after Sally relented and let the girl come to visit, Tuck kept his distance. Now he was paying for it. *Guilt could destroy a man,* he thought, swishing his beer around. *So could grief and anger. Or the need for revenge.*

With one swift tug of his beer, he dismissed such thoughts. A beer could drown anything out, couldn't it? Even that hollow, silent cry trapped in his throat.

"Yo, Wheaties," he yelped as the lieutenant posed at the edge of the diving board. "Keep your trunks on this time."

Wheaties grinned, then hollered his trademark call: "Hey Drivers. It's live to drive and drive to live."

Beating his fist against his chest, and yelling like Tarzan, Wheaties threw his muscular body into a powerful spin, somersaulting twice in the air before plunging into the water.

Buzz Hawkins walked up behind Tuck. "That looks like fun, Colonel. It's been years since I've done my world-famous cannonball. Think I'll join them."

"Clear the deck," Jo-Ellen roared.

She and Gina slipped into the house to give Krystal a hand in the kitchen.

Tony uncovered the grill. "You like the livers?"

Tuck turned around. "Say what?"

"Chicken livers." Tony tested a piece of chicken with a fork. "They're my favorite part of the bird."

"I like the wings," Tuck craned his neck skyward. There wasn't a cloud in sight. "Good day for flying."

Tony looked up from the grill, surrounded by a cloud of smoke. Sweat trickled down his dark temples; below him, the chicken sizzled over a bed of coals. "Good day for a barbecue, too." He whipped out a hanky to mop his face.

Tuck took a swig of beer. "Sure smells good."

Tony stuffed the hanky in his back pocket. "Pass me that platter, would you, Colonel? Before I burn dinner. Krystal says there's nothing worse than sinking your teeth into a piece of dried-out, rubber chicken."

***

In downtown Bolton, on the banks of the Red River, a large crowd gathered before dusk to watch the annual fireworks display sponsored by the city.

At the center of a makeshift stage, embellished with red, white, and blue rosettes, Beauregard's wing commander, Colonel Glen P. Dennison, started his annual Fourth-of-July speech.

In a folding chair next to the mayor, in the front row of VIP seating, Lois Dennison waited in a linen suit and low-heel pumps. She folded her hands in her lap and listened dutifully to the same old speech she could have delivered herself.

She stifled a yawn. She couldn't wait to get home and shuck off her blasted pantyhose.

To her immediate right, and a tad too close for comfort, sat the ever-attentive Captain Linda Garrett, the blonde shadow who followed Lois' husband everywhere. Captain Garrett's face turned to admire her boss. *So what else is new,* Lois thought grimly.

Across the river, directly behind her husband, the paper mill belched away.

Chapter Nine

Sunday morning, July 8, Sylvia Wheaton pulled the Jeep into the parking lot of the base chapel and glanced at her husband. "Are you nervous?"

Wheaties looked up from a shiny brass trumpet he held in his hands. "Heck no." He limbered his fingers up and down on the instrument's three valves.

Sylvia laughed. "You're a regular ol' ham at this, aren't you?" She navigated the Jeep into a parking slot in front of an enormous stained glass window that depicted a dove and an airplane flying skyward toward a brilliant sunburst.

Wheaties turned the bell end of his horn toward her and blew a couple of practice notes before he stuck the trumpet back in its case.

"You're a mess," she razzed him and dropped the keys into her purse.

"That's why you married me, Darlin'." He snapped the case shut and flashed a grin.

They met at Kansas State when Wheaties, the football jock and amateur musician, charged into the campus library late one night to cram for exams. Sylvia, studying at the next table, had to ask him twice to stop cracking his knuckles. On the second try, Wheaties picked up his books, moved to her table, plopped down in front of her. "What do you have against redheads?"

After that night, Wheaties' fraternity brothers, most of them football players, couldn't figure out why one of the most popular guys on campus suddenly dropped the flashy sorority chicks for a shy, fresh-faced honor student straight off a Kansas farm. With a wink, Wheaties quipped, "Because we make great music together."

Sylvia tutored the party boy back to straight As, and he played his trumpet for the brainy girl who couldn't carry a tune. She brought out the thinker in him buried beneath muscle and tendons while he made her laugh with his zany antics. One night, after getting injured in a football game, Wheaties told her about coming home from school when he was a kid and being told his daddy was dead. For the first time in his life, the tough-guy football player cried in front of a girl.

"I'll sit with Gina and the munchkins," Sylvia said. "I hope Michelle comes."

"You two sure hit it off at the party the other day," Wheaties nodded.

"I know what it feels like to be an outsider," Sylvia shrugged.

Wheaties gazed at her thoughtfully. "I'd be sprayin' crops right now if it weren't for you. I'd have flunked out of college and blown my chances to fly fighters."

"You would have buckled down and studied," Sylvia said. "You wanted it too badly."

Wheaties leaned back against the headrest. "You're the best thing that ever happened to me." He turned to look at her.

Sylvia blushed. She reached over and touched her husband on the arm. "Any chance Colonel Westerfield'll show up? He's never heard you play."

"He doesn't do church, Darlin'," Wheaties chuckled and went to get out of the Jeep.

"Why was he grounded?" Sylvia asked. "Nobody's talking."

Wheaties glanced at his watch. "You know as much as I do. He hurt his back. Church starts in half an hour. We better scoot. Airman Crenshaw is probably wondering what happened to me."

"You don't believe it either, do you?" Sylvia frowned.

"I get paid to fly airplanes, Darlin', not ask questions." Wheaties grabbed his trumpet case and headed up the sidewalk.

Sylvia lingered, blotted her face with a tissue. Suddenly, she felt ill. She hopped down from the Jeep and followed Wheaties up the sidewalk.

Halfway there, she stopped abruptly and put her hand to her mouth.

Wheaties started back towards her. "What's wrong?"

"I feel like I'm going to be sick."

"Here, let's get you inside where it's cooler."

He towered her by a foot and parked her in a chair in a hallway off the main sanctuary.

"How ya feeling?" He bent toward her like any other serious young professional in a charcoal-gray suit and polished shoes. But serious stopped at the tie dangling down the front of his crisp, white dress shirt. The tie featured a circus clown perched on a unicycle, his thumb in the air. The clown had a bandage taped to his bulbous nose.

Sylvia grabbed him by the tie and reeled him closer. "That is the ugliest tie I've ever seen in my life."

Wheaties chin dropped. "You don't like it?" He flipped it over and revealed a naked lady. "Uncle Matthew gave me this tie when I graduated from pilot school."

Sylvia fumbled with the tie. "You need a tie clip, Honey. What if Chaplain Lawrence sees it?"

Wheaties shrugged. "He's a good shit. So how's my girl?"

She leaned her head against the wall, looking pale.

Wheaties felt her forehead. "You feel cool as a cucumber."

People started to arrive.

She nudged him away. "It's getting late. Airman Crenshaw's probably back in the choir room thinking you stood him up. I'll see you in church."

"You sure?" Wheaties hesitated.

"Scat, trumpeter. Go blow that horn of yours."

After Wheaties picked up his trumpet case and walked off, Sylvia slipped into the ladies' room off the narthex and slumped into a chair, praying for the nausea to pass. It was then she remembered the weekend they spent in New Orleans, when she forgot to take her birth-control pills.

***

Gina cracked open the bathroom door and poked her head inside. "If your daughter thinks she's wearing those boots to church, she's got another think coming. And on top of that, her nipples poke through her vest."

Tuck rattled the Sunday newspaper. "I'll be there in a minute."

Gina started to close the door. "At least she's not wearing that white gunk on her face."

"That's a start," Tuck replied behind the black-and-white newsprint.

"Wheaties is playing his trumpet. Why don't you join us?"

A single eyebrow shot up over the corner of the paper. "Tell Wheaties hi for me." Tuck snapped the paper. "Now do you mind, Babe?"

Gina pulled the door behind her and padded off across the bedroom and grabbed her sandals from the closet. On her way out, she caught her reflection in the mirror over the dresser.

With her dark hair piled on top of her head, and a lacy collar framing her slender neck, she looked like the picture of serenity, like a cameo on an antique brooch.

Her appearance gave the illusion that everything was hunky-dory. If she could convince others, then maybe she could convince herself.

In the kitchen, she unplugged the coffee pot and dumped the grinds in the sink. The boys squealed in the other room.

Irritated, she grabbed her purse and sailed into the living room. Michelle had the boys pinned to the floor, tickling them. The movie *Top Gun* played backward on the TV screen. Jesse had seen the movie about a plucky young Navy pilot at least half a dozen times since Tuck bought the video.

"Everybody up," Gina ordered, marching over to the VCR. "Church starts in fifteen minutes."

Michelle scrambled to her knees and tucked Jesse's shirttail in. "Hold still, Squirt."

Gina pressed the eject button and the tape popped out. "Austin, tie your shoes." She slid the video back into its jacket and stashed it in a cabinet under the TV. "Jesse, don't mess with this VCR until I teach you the proper way to rewind a tape."

"Isn't Lieutenant Wheaton playing his trumpet today?" Michelle asked.

"Yes, and if we don't hurry, all the front pews will be taken." Gina tapped her foot, glanced over her shoulder, and waited for Tuck to get out of the bathroom.

"My old boyfriend use to play the trumpet," Michelle snapped Jesse's pants and bent to help Austin tie his shoes. "Until he got his front teeth knocked out in a fight."

Gina checked her watch again, half-listening.

"He turned out to be a real dud," Michelle went on. "He sat last chair for three years in a row and never did learn how to read music. The only reason he joined the band was to go on the band trips."

"So why did he get in a fight?" Gina asked.

Michelle rose to her full height, three inches taller than Gina. "Some dude called him a band pixy."

Gina burst into a laugh.

Tuck marched into the room with his morning Coke and the paper. Dressed in khaki shorts and a T-shirt, and smelling faintly of aftershave, Tuck tossed the paper aside.

Gina waited for him to say something. Instead he stood with this dumb look on his face while his eyes swept over his daughter.

Gina couldn't blame him. Despite the grim reaper eyeliner and lipstick, and the choppy haircut, Tuck's daughter was beautiful. She had high, prominent cheek bones, a smooth, perfect nose and a sexy dimple planted in the middle of a firm, solid chin. The micro-miniskirt and a simple white vest only enhanced her slender figure and voluptuous bustline.

Tuck took a sip of Coke, cleared his throat, then turned to address his daughter.

"Sweetheart, don't you have some sandals or something?" He took another sip of Coke, avoiding the other issue.

Michelle glanced at her feet. "What's wrong with my boots?"

"Nothing, really," Tuck grunted. "I just think some other shoes would be more appropriate for church."

"How would you know what people wear to church? You never go."

Gina gnawed her lip as Tuck's face fell, then hardened, as if the comment hadn't fazed him.

"Tell you what," Gina said in a sudden burst of cheerfulness. "Let's compromise, shall we? Michelle, you can wear those boots till the soles rot out. Under one condition."

"What's that?" Michelle rolled her eyes.

Gina lowered her voice. "You have to wear a bra to church."

Austin giggled and covered his mouth.

Jesse yanked at Michelle's skirt. "Mommy gots a bra,"

he said. "She wears it on her boobies."

Michelle blushed. "I didn't bring one," she said, meekly. "Got one I can borrow?"

Gina hesitated. "Honey, I don't think mine would fit you."

Michelle glanced down at her bustline, then looked over at Tuck.

Tuck shrugged. "Hey, don't look at me. I can't help."

Gina racked her brain. "I have an idea, but we'll have to hustle. Tuck, you load the boys in the truck while we run next door."

"Next door?" Tuck gave Gina one of his you gotta be kiddin' me looks.

Gina ignored him and swept out of the house with Michelle in tow.

"When did you stop wearing a bra?" Gina asked as they cut across the front yard.

"When Mom made me quit my job." Michelle side-stepped a fresh mound of fire ants. "I hate those cocksuckers," she stomped her boots on the grass.

Gina threw her a disapproving glance. "You sure use that word a lot. Isn't that how you described your neighbors in Galveston?"

Michelle grinned sheepishly as they came up the driveway. "What, cocksuckers? Yeah. Some of them are okay. It's the lazy bums on welfare I can't stand. They're always coming around the apartment asking Mom for handouts."

Gina rang the doorbell. "Oh, really?"

"Yeah," Michelle said, checking her feet for stray ants. "Mom says you never know, any one of 'em could be Jesus."

Gina cleared her throat. "What does your mom say about the way you dress? Surely she doesn't approve."

"She hates everything I do," Michelle said with a dry laugh. "She always talks to God about me. She does this thing called laying on hands, where she squeezes my shoulders and chants her gobbledygook and spouts her scriptures." She cocked her head to the side and looked at Gina. "Mom's crazy, you know. She says my soul's been kidnapped by demons."

Wynonna Sandford flinched when she opened the door. She cuddled her dog closer, like an overprotective mother wary of some communicable disease going around.

Gina put her hand up in a peace offering. "Look, Wynonna. Whatever happened the other day, that's between you and Tuck."

Wynonna hesitated. She gawked up and down at Michelle, then invited them in.

After Gina explained their predicament, Wynonna trotted off in spiked mules to fetch one of her old brassieres from the days when she was twenty pounds lighter.

From somewhere in the back of the house, Big Sandy blustered, "Schnookums, I've heard of neighbors borrowin' sugar and eggs. But this takes the cake."

Gina and Michelle giggled.

When Wynonna returned, minus her dog, she had a big smile on her plump, painted face.

"This thing's bullet proof," she passed the stiff-coned 36DD bra to Michelle. "It might itch a little but it should do you for church."

Michelle crinkled her nose. "Thank you," she dangled the girdle-white thing in front of her like a string of stinky fish.

"I owe you one," Gina said at the door.

Wynonna batted her lashes impatiently at Gina, then turned to Michelle.

"Now listen here, Angel. There's nothing wrong with being proud of what the Good Lord gave ya, but take it from an old pro like myself." She paused and patted herself on the chest. "You need to keep those things in a holster, Honey. They're deadly weapons."

Wynonna reached up on her tippy-toes and pinched Michelle on the chin. "You come by here anytime, Angel." She moved her head from side to side and examined the girl's face. "I'm just dyin' to give you a make-over."

***

Tuck was alone with the sound of the antique clock ticking against the gray, lifeless silence that settled over the house when Gina and the kids left for church. No screech of boys, garbled television voices, phones, or loud rock music to jangle his nerves. He couldn't bear the silence.

He went outside and stood on the patio, feeling restless and edgy. The boys' green turtle sandbox smiled back at him from the corner of the patio. Fire ant mounds sprung up right and left all over the yard. No sooner Tuck killed one mound than another turned up five feet away. To knock off the queen was a bitch.

The grass needed cutting, the bushes trimmed. Any day he would get a nasty-gram from the housing office to cut his yard. So many restrictions. So many rules. So much bullshit.

Austin's metal baseball bat leaned against the side of the house. Tuck picked it up and swung it a few times. Baseball was never his favorite pursuit as a kid. When he wasn't fishing, he built model airplanes or read books about airplanes. By Austin's age, Tuck had his future planned. He pulled up a lawn chair, sat down, and held the silver bat.

The warm metal felt good in his hands. He switched the bat to his right hand, then propped it between his legs, the barrel end braced against the concrete between his feet. His right hand formed a fist around the handle. Tighter and tighter Tuck squeezed the bat, maneuvering it forward and aft, left and right, like the control stick of a fighter jet.

He missed the feel of the stick in his hand, his feet on the rudder pedals, the intoxicating smell of jet fuel and the sheer power of the A-10's engines roar to life while he strapped himself in, all snug in his titanium bathtub, his cockpit.

Yet he wanted out of the Air Force. Jeff Sweeney had wanted out, too. Away from the hassle, away from the dogfights on the ground. Pilots screwed their buddies and jockeyed for position to wheel and deal themselves into the best cockpit assignments or desk jobs to boost them one step up the chain of command. Everybody kissing everybody's ass to get promoted.

"They don't kiss ass in the airlines," Sweenedog assured him. "Promotions are based on seniority, not politics."

They don't kill their pilots on a regular basis either, Tuck thought and gripped the bat tighter and tighter, till his hands hurt. Weeks of anger and rage, and years of pent-up grief, boiled up as he sat in the lawn chair, squeezing the bat so hard it started to shake. He felt as though he was a hundred feet above the earth, in an out-of-control-aircraft, heading straight for the ground, and all he could do was strangle the control stick as the forces of gravity had their way.

In reality, he strained against forces he could no longer control. But this time, when the anger, rage, and grief came to a head, he didn't reach for a beer to drown it. Or shrug it off in indifference and walk away.

He exploded in a fit of rage and flung the bat across the patio. It put a dent in the side of the plastic sandbox before it bounced onto the concrete in a terrible clatter.

Tuck heard somebody gasp. He looked up. Standing in the middle of her yard, Wynonna Sandford shook her head at him in disgust. She jerked Baby by the leash, made a face at Tuck, and scurried inside.

Tuck walked over to inspect the damage. One side of the green turtle was slightly caved in. At least he missed the head.

He stooped to pick up the bat.

It felt good to let off steam. Time to stop pussyfooting around and put in his papers.

He knew what Sweenedog would say. "Shit or get off the pot, Westerfield."

Tomorrow, Sweenedog. Tomorrow, Tuck thought. *Bright and early, when they least expect it, I'll march over to base personnel and put in my retirement papers. I'll do it for both of us.*

Ah, Jeff Sweeney, cocky and blond, full of energy, and everybody tried to emulate him back then.

When Sweenedog was the first one in their class picked to fly solo, he turned to Tuck and the others, before strutting out to his jet, and yodeled good-naturedly, " It's show time, boys."

Now he was dead.

Tuck propped the bat against the house and went inside. He got a drink of water at the kitchen sink when Wynonna's purple Suburban tore out of her driveway and zoomed off down the street.

The silence in the house crept up on him at once, as if an invisible presence stood back watching him with critical eyes. *I need to do something. Cut the yard. Slaughter more fire ants. Go to church.* "Wheaties is playing his trumpet," Gina said earlier.

He glanced at his Rolex. *Church is halfway over. I can't walk in late, can I?*

"Better late than never," Granny's voice chirped in his head, her weekly proverb when he was a boy, as she roused her reluctant clan from the breakfast table in a flurry to get to church.

"Church is full of hypocrites, Belle. Why bother?" Tuck's mother would complain, while the rest of the family hustled out the door.

Tuck's grandmother would pat him on the back and chuckle, "If you can't beat 'em join 'em, I always say."

Tuck put his glass in the sink and went to change. The yellow Polo and tan Dockers would have to do. The last suit he owned was twenty years old, and Gina said it made him look like a lounge lizard. *It's show time, boys.*

***

Wynonna scowled at her dagger-length fingernails fanned over the top of the steering wheel like purple claws. The customized Suburban va-roomed down the four-lane stretch of highway as if on autopilot. Wynonna made this trip to Bolton every Sunday morning. Some folks went to church to find God. Wynonna went to a noisy, smelly building on the other side of town to save lost souls.

"I'll have'ta redo my nails as soon as I get home," she fretted to herself. "A Purple Passion beauty consultant doesn't traipse around with chipped nail polish. What would my customers think?"

Customers. She had oodles of them. All over base, in town and practically the whole town of Rose Glen. She struck the mother lode when she set up a deal to donate a percentage of her profits to the Rose Glen Gators Booster Club in exchange for the exclusive rights to hawk her beauty and skin care products at all their home games. After that first football game, she had ladies calling out the kazoo, asking for more Purple Passion.

Wynonna ran a mental checklist of the numerous orders she had yet to fill and deliver. She would do the base tonight and hit the streets of Bolton first thing in the morning. Thankfully, she didn't have to go back to Rose Glen for another week, after spending an entire afternoon there last Monday. So much to do. So little time.

"That's the problem when you're good at your job," she sniffed. "Customers come to expect your prompt and professional service every time. And if you don't give it, the competition will."

She flipped down the sun visor to inspect her roots. "Good gravy." She snapped it shut, seized the steering wheel and concentrated on her driving.

A few minutes later, she barreled into the empty parking lot of the city dog pound and screeched to a stop. She fretted again on how to find time to redo her nails, touch up her roots, throw a pot roast together, and still make her deliveries before supper time.

"Stop throwing yourself a pity party," she scolded. "You didn't get this far in life cuz ya sat around lookin' purty all day."

Getting out of the Suburban in tight jeans took some doing, but Wynonna had it down to a science. Wriggling to the edge of her seat, she pushed the driver's door open, swung both legs out to the side, then hopped down on her stiletto heels, using her hip to close the door.

She slipped a key in the lock of the building and let herself in, greeted by a chorus of barking dogs and mewing kitty-cats. Their weekend savior was used to the stench of caged animals, but she could never get used to the cold cement floor or the eye-stinging ammonia the regular workers hosed it down with.

"I'm coming, my babies," she cried with compassion and tottered over to a tall counter to drop off her purse. Nothing else mattered but to coddle and make a fuss over a slammer-full of sad-eyed faces awaiting homes. Or eternal rest, if they weren't adopted soon.

"I'd take you all home with me this minute," she informed them as she walked up and down the center aisle, peering into their cages. "But the Air Force don't take too kindly to four legged critters."

She stopped by one cage where a black Doberman pranced nervously and whined.

"Hello, Prince Charming. Mama's sorry she was late today." The Doberman came over, sniffed her, then settled down.

A tiny Chihuahua, not to be outdone by the bigger dogs, yapped her hello.

"Ah, Senorita Chee-Chee," Wynonna hunkered down in front of her cage. "How's my little Mexican jumping bean today?"

Straightening back up, Wynonna took an inventory of how many mouths she needed to feed before continuing on with her explanation of why she was late.

"Being it's a Sunday and all, I had to be a good Samaritan and loan one of my old brassieres to a neighbor's step-daughter," she chuckled. "Daddy sure thought it was a strange request, but what does he know about girls' stuff?"

She sashayed back up to the front to get the food, giggled with pleasure and slashed open giant bags of dry dog food and cat food.

A black cat batted at something in his cage, and Wynonna looked up to watch it.

"What'cha got there, Bad Luck?" She slit open another bag.

Bad Luck went right on pawing at the air and Wynonna went right on talking like he listened. "Imagine that," she prattled. "This girl Michelle gallivants around in men's combat boots without a bra. What's the world coming too when a purty girl like that messes up her face trying to look ugly? Granny and Pawpaw would'a hauled me out to the woodshed."

She shook her head in bewilderment and loaded the food on a big steel cart she rolled down the center aisle, like a flight attendant doling out peanuts and drinks.

"Here's a big bowl of crunchies for my friend Goldie," she said to a Golden retriever who wagged his tail. She moved to the next stall.

"Hey, Sarge." A hollow-eyed German shepherd, his ribs sticking out, lifted his head to look at her while she poured food in his dish.

"I felt like slamming the door in Gina's face," she confessed into his sad brown eyes. "Till I got to figuring, why blame Gina? It's not her fault she married such an ass. Whoops," she giggled, and pushed her cart along. "'Scuz my French. Now you eat up, Sarge. You're startin' to look like a bag of soup bones."

When she got to the end of the aisle, she stooped to pick up a small yellow tabby with mangy fur. "You're new, aren't you, Angel?" Stroking the cat's head, she cooed, "There, there." The cat blinked and purred with pleasure, cradled in her arms.

Wynonna stood a long time, rocking the cat back and forth and hummed *Rock-A-Bye-Baby* while tears welled up in her eyes. "It'll be okay," she whispered in a raggedy voice. "Mama's here to take care of you. A long time ago, when I was no bigger than a kitten, I wasn't wanted either."

***

Chaplain John Lawrence, the head Protestant chaplain, led the congregation in prayer. Tuck ducked into a squeaky pew at the back of the chapel. He spotted Michelle in the second pew from the front, nestled between Gina and Sylvia Wheaton. Tuck bowed his head and felt strangely disappointed at the empty spaces in the pew beside Gina.

Austin and Jesse were at children's church. Tuck pictured them stuck in some back room of the chapel, coloring pictures of Noah's ark and Moses parting the Red Sea. Or was it the Dead Sea? Tuck couldn't remember. It had been too long.

"By Our Lord and Savior Jesus Christ..." Chaplain Lawrence's voice resonated from the pulpit.

Tuck winced. *Jesus. Now there was a name Tuck used often and with fervor. Jesus H. Christ. Sweet Jesus. So help me, Jesus. For the love of Jeee-zus.*

Needles of guilt pricked his conscience.

I see you're still taking the Lord's name in vain, Sally lambasted him on the phone.

Tuck shifted in the padded pew as Sally's self-righteous accusations rattled around in his head from a lifetime ago. *You're a sinner, Tuck. A heathen. A baby killer. Repent from your evil ways and get out of the Air Force, or I want a divorce!*

"Amen," somebody shouted.

Tuck's head snapped up. The prayer ended.

The chaplain invited them to open their hymnals to page 375, for the closing hymn. Or was it page 275? Tuck missed it.

"All rise," said Chaplain Lawrence with a benevolent smile as the congregation stood up to sing, *Fairest Lord Jesus*.

Tuck fumbled with the hymnal as he scrambled to his feet, peeked over the person's shoulder in front of him for the right page number – 275.

He rifled through the silky thin pages for *Fairest Lord Jesus*. When had he lost touch with his childhood Jesus...that brown-eyed, bearded fellow hanging from the wall above his bunk bed when he was a kid, that gentle friend in the eight-by-ten he talked to at night after his mother tried to dissuade him from becoming a pilot.

Was it college, when pictures of sexy airplanes and centerfolds took Jesus' place on the wall? Or pilot school, when Tuck got so good at compartmentalizing, at camouflaging his feelings. Or the year in Vietnam, when he decided he no longer needed to confide in anyone, much less some entity he couldn't see or touch.

Midway through the second verse, singing something about Jesus being purer, Tuck lost his place. He found it far more entertaining to watch Chaplain Lawrence at the front of the sanctuary sing his heart out, than to listen to his own dull voice trying to blend in with those around him.

Chaplain Lawrence was tall and slender, with kind, gentle eyes and a benevolent face. For a man of his maturity and rank—a Lieutenant Colonel like Tuck—the chaplain had an almost childlike quality that Tuck found refreshing. They met before, at wing staff meetings and such, but strictly business.

John Lawrence was a nice guy for a preacher, Tuck decided, as the congregation moved into verse three, something about sunshine.

Preachers. Tuck never trusted the lot of them. Like the Southern Baptist preachers at his grandmother's church in Rose Glen. Tuck decided that's when he lost touch with his childhood Jesus. When he started confusing Jesus with all those plainclothes dragons that spit out hell, fire, and brimstone, instead of love and forgiveness.

When the hymn ended, Chaplain Lawrence waited for everyone to sit down before he continued.

"Instead of the usual benediction," he explained, "we'll close with a special musical presentation." He introduced the soloist, an airman straight out of basic training, named Brian Crenshaw. A short fellow with a patch of dark hair left the choir loft and made his way to the microphone.

"Airman Crenshaw has been assigned to the base photo lab," Chaplain Lawrence informed the congregation. "And we all know Lieutenant Wheaton," the chaplain chuckled. Wheaties rose from his seat and came forward, fiddling with his tie.

Wheaties lifted the trumpet. Halfway to his mouth, he hesitated, and grinned like a little kid in a school play who'd just spotted his mom or dad out in the audience.

Tuck gulped, wanting to slide under the pew. He acknowledged the lieutenant with a nod.

Then with military precision, Wheaties pulled the trumpet to his lips, drew in a deep breath, and played the opening notes of *The Lord's Prayer.*

The moment the music hit the air, Tuck's throat ached. The melody was so sweet and tender it almost hurt. The music flowed out of the mouth of the trumpet and poured into Tuck's soul.

Caught off guard, he was unprepared for the affect it had on him.

When the soloist joined in singing the opening stanza,

"Our Father, which art in heaven..." Tuck closed his eyes, giving in to its power.

Moments later, his eyes fluttered open as the soloist sang "...as we forgive those who trespass against us..."

To his surprise, Tuck found himself staring at the back of Colonel Dennison's head several pews up. The colonel had his arm draped over the back of the pew. Mrs. Dennison sat next to him at a respectable distance, with enough room left over to squeeze in another person.

That person, Tuck realized, could have easily been Captain Linda Garrett, looking as peppy and pure as ever, who happened to sit in the pew directly behind her boss, along with her husband, Hank, and the couple's two kids.

Any closer, and she could have blown on Dennison's neck.

Well, Tuck thought, *Mother was right about one thing. Church is full of hypocrites.*

Airman Crenshaw hammered out "...for thine is the kingdom..."

Tuck gazed at the stained glass window where the dove and the airplane flew into the sun.

His decision was final. Tomorrow was D-day. The day he would confront Dennison and take control of his own destiny. But the only way to do that, and to do it successfully, was to get up at the end of the service and walk out of church, and leave his anger at the door.

When Airman Crenshaw belted out "...and the power..." Sylvia Wheaton bolted down the aisle, as white as a ghost, her hand clamped to her mouth.

She shot past Tuck as the music built to a crescendo: "... and the glory..."

Tuck sprang to his feet and went after her.

Halfway out the door, just as Airman Crenshaw and Wheaties hit the first notes in "...forever..." Sylvia threw up.

***

After church Tuck stopped by the shoppette, a small convenience store on base. He wedged a bottle of Scope between his chest and a large bag of ant poison, and headed for the register. As he rounded a corner by the shampoo and soaps, he nearly bumped into a scrawny, freckle-faced security policeman dressed in camouflage fatigues and a blue beret.

Sergeant Willis Chambers flinched. "Oh, excuse me, Sir." The sergeant jerked his shopping cart to keep from hitting Tuck. Tuck hadn't seen the sky cop since the night at the stables.

"Hey, Sergeant Chambers. How's it going?" Tuck reached for a bar of Irish Spring.

The sergeant shifted from one foot to the other. "Can't complain, Sir." He couldn't seem to look Tuck in the eye, not after holding him at gun point. "I'm on lunch break," he said, as if he needed to explain himself. "The wife needed milk and diapers."

Tuck felt sorry for the young sergeant, who bore the look of a man who made a pact with the devil. Until two weeks ago, Sergeant Chambers pulled the graveyard shift, four to midnight. Now, for the first time in his ten-year career, he had a "normal" schedule, eight to four, Monday through Friday, working alternate Sundays. Courtesy of the wing commander.

Job security, medical benefits, and a chance to see the world—that's why Sergeant Chambers agreed to keep his mouth shut about the cover-up. Tuck couldn't blame him.

The sergeant lowered his voice. "Sorry about the other night, Sir." Beads of perspiration broke out above his upper lip.

Tuck narrowed his eyes. "Look, Chambers. You saw a drunk weaving down the middle of the road on foot. You were just doing your job."

The sergeant stood a little taller. "Thank you, Sir."

"Besides," Tuck quipped, "If I tried to drive home from

the club that night, you would have nailed me with a DWI. You're a good sky cop, Chambers. You did the right thing."

The sergeant blushed.

"Well, I better be going, Sir." He glanced at the bag of poison in Tuck's arms. "You've got fire ants, too, Sir?"

"Doesn't everybody?" Tuck chuckled and walked off.

Chapter Ten

"Westerfield put in his papers." The chief of personnel's voice went off like an air raid siren in Bull Spitz's ear.

"Retirement?" The squadron commander put the phone down and shut the door to his office. Back at his desk, he crammed a fresh stick of Juicy Fruit in his mouth and picked up the phone. "Are you absolutely certain?"

"Westerfield just waltzed out the door."

Bull Spitz chewed on his gum, trying to digest this unexpected change in events. They hadn't planned on Westerfield retiring, not with his name on the squadron commander's list. It didn't make sense. "When's it effective?"

"Next spring."

To Spitz disgust, he found himself allied with the biggest laughingstock on the base. All because he accepted a ride home from the sniveling ground pounder who lived next door. Now both of them tried to cover Colonel Dennison's hide.

"Westerfield has something up his sleeve," Vic Rollings added in a panicked voice.

"Rollings, you'd never make it as a pilot. You'd pull the ejection handle at the first sign of trouble."

"Very funny," Rollings sniffed. "But let's remember who was too stoned to drive home from the club the other night."

Bull Spitz stopped chewing his gum. "And I appreciate the lift." His chair squeaked as he rocked back and forth against the wall. "But let's not forget whose bright idea it was to check out the flash of light behind the barn. Some UFO, huh, Vic? Now don't get your knickers in a knot. You'll make full colonel out of this."

"Only if Westerfield continues to cooperate. He's got nothing to lose if he blows the whistle now, except his privacy."

Spitz stopped rocking and sat straight up in his chair. Rollings had a point. "I'll inform the boss."

***

"Spitz was right, Marsh," the wing commander reported to his boss. "Westerfield's a loose cannon."

"Not to worry, Glen," the lieutenant general said. "He can't retire until you sign those papers. Persuade him to change his mind. Make him a deal so sweet he'd be crazy to refuse."

"What do you propose?"

"Give him command of the 428th!" Honeycutt ordered. "Effective next spring, when Spitz turns over the squadron."

"Let me warn you, Marsh. Westerfield's not much of a politician."

"That's the trouble with today's Air Force," the general thundered. "Too many goddamned politicians. At least Westerfield has combat time—something Spitz doesn't." The general clicked off.

Dennison got Bull Spitz back on the line. "Drop whatever you're doing and hustle over here. We're going to buy him off."

"Don't tell me," grumbled the squadron commander. "The 428th?"

"Take it up with the general," Dennison snapped and hung up.

Ten minutes later, Bull Spitz parked himself on a leather couch in the outer suite of the wing commander's office and flipped through a current issue of *Air Force Times*. Every other page or so, he glanced at his watch, then at the door.

Colonel Dennison's secretary, Lucille, otherwise known as The Palace Guard, bent over her computer, squinted at something on the screen. Her bifocals rested above the tip of her nose.

The door to Colonel Dennison's office sprung open and out stepped Captain Linda Garrett with a maroon Geodex dayplanner and looking as youthful and efficient as ever.

"Have a good day," the diminutive captain tooted and zipped past Lucille like a Volkswagen Rabbit, blowing past an old, clunky Oldsmobile.

Lucille frowned. "Same to you," she muttered, and peered over the rim of her bifocals.

Colonel Dennison's voice buzzed over the intercom. "Any sign of Lieutenant Colonel Westerfield?"

Lucille mashed the transmit button. "No, Sir," she replied, just as the sound of coins rattled in someone's pocket in the hallway. Lucille looked up expectantly. So did Bull Spitz.

"Take that back, Sir. I think that's him now."

Tuck appeared in the doorway, a dark-blue folder in his hand. He greeted Lucille with a warm smile. "Any chance I can get in to see the wing commander?" Tuck glanced at Spitz, hiding behind *Air Force Times*.

The secretary's eyes softened. "I believe Colonel Dennison expects you. One moment please."

She buzzed the wing commander.

"Send him in," Dennison's voice crackled over the intercom.

***

Dennison's brows snapped together. "What do you mean, you have other plans?" He leaned forward across his massive desk. "I offered you the chance of a lifetime."

Tuck sat as stiff-spined as the Lincoln Memorial. "It's time I moved on," he explained. "My boys need roots. I don't want them changing schools every two years because the Air Force says it's time to move."

Out the corner of one eye, Tuck could see Bull Spitz smack his gum next to him. Each time Spitz's large, carnivorous jaws opened and shut, the vein in the middle of his forehead pulsated to life.

Dennison cleared his throat. "Now Tuck, let's be reasonable. You're not a captain anymore, or a major. You're a senior officer with a bright future. As far as your children are concerned, military brats get to see the world. Hey, it's one big field trip," he observed with a chuckle. "Stick around, Tuck. I'll have you back on flying status within the hour. You'll be a squadron commander this same time next year."

"Thank you, Sir, but my decision is final." Without any fanfare, Tuck opened the dark-blue folder, extracted a piece of paper and slid it across the desk.

Even upside down, Tuck could read the large, bold print at the top of the form: "Application for Voluntary Retirement from Active Duty Service in the United States Air Force."

A sense of freedom rushed over him. In nine months, Tuck would be a civilian. No more rank and structure and all the mandatory "Yes Sir-No Sir-Three Bags Full Sir" acknowledgments he made to his superiors over the years.

"Put that away." Dennison barely looked at the form. "If you find the 428th unsuitable, you can have your pick of assignments. There's a squadron opening up at Myrtle Beach. I can arrange it."

"Sir, as my daddy used to say, 'I've ridden this old horse as far as she'll take me. It's time to put her out to pasture.'"

Dennison blanched. "This may be the best offer you ever get."

"The boss is right," Spitz interjected smugly. "Take it from me, Tuck, there's nothing better than to command your own squadron." He puffed up his chest. "It's the greatest achievement of a fighter pilot's career."

"What about combat?" Tuck shot back, catching both men off guard. "Doesn't combat experience count for much these days?"

A hush of silence fell over the room. Spitz glared at Tuck, then looked away.

Finally the wing commander spoke up. Once more he tried to persuade Tuck to accept his offer. And once more Tuck declined.

Dennison's eyes narrowed. "Tuck, you're making the biggest mistake of your career," he barely moved his lips. He picked up his pen, scribbled his signature on the form, and slid it across the desk.

Tuck slipped it into his folder, stood at attention and waited to be dismissed.

"That's all," Dennison said through clenched teeth.

Like a good soldier, Tuck saluted sharply, executed a perfect about-face, and marched from the room. He hummed *The Old Gray Mare She Ain't What She Used To Be* on his way out the door.

Out in the hall, Tuck heard Dennison bark, "Okay, smart guy. What's Plan B?"

Back in his own office, Tuck told Sergeant Duran to take the rest of the day off. Clutching the blue folder, Tuck strolled down the hallway to a pay phone and called Gina.

"You sound out of breath," he piped.

"We just got back from the mall. We're leaving for the pool as soon as the kids change." Gina paused. "Why don't you meet us for lunch? We can picnic under the…"

Tuck cut her off. "I'm on my way home. We need to talk."

"About what?"

"About growing old together in a real house with a big front porch and a fireplace where the kids can hang their Christmas stockings."

"Tuck, quit beating around the bush. What are you getting at?"

"We'll talk about it when I get there. Let Michelle take the boys to the pool. She might as well take the Bronco."

"I don't know, Honey. I mean, anything could happen."

"Gina, we have to start trusting her sometime. Just tell her to be careful. What were y'all doing at the mall?"

"Buying Michelle a bikini."

"You're a bad influence," he teased and hung up.

***

"For Christ's sake. She's a certified lifeguard," Tuck said.

Gina pictured Jesse's lifeless body at the bottom of the O'Club swimming pool. She sat on the edge of the living room couch and watched Tuck pace in front of her.

Except for the tick of the antique clock on the shelf and rattle of loose change in Tuck's flight suit, Gina heard little else but the sound of her own breathing, in-out, in-out, like the sound of a respirator. Another image flashed by in her mind, this time of Jesse hooked up to a life support system, his brain damaged from lack of oxygen.

She blinked, trying to dismiss the mental picture, only to find a new one moved in to replace it. This time Austin's wiry, wet body scrambled up the steel rungs of the high dive, then slipped and fell backwards onto the concrete.

"The boys are in good hands." Tuck brushed past the

couch. When he reached the end of the room, he turned and retraced his steps.

"You're going to wear out the carpet," Gina said, annoyed that Tuck could be so blasé about letting the boys go off to a crowded swimming pool with Michelle. At the same time, she had a knot in her stomach, dreaded whatever news Tuck worked up the courage to tell her.

Tuck halted mid-stride and turned to face her. Gina braced herself.

"I put in my papers," he said. "I just came back from the personnel office."

Gina leaned forward. "You don't mean *retirement?"* She blinked in astonishment.

"It's time to press on. Pursue other dreams."

"But the Air Force is your life."

"Maybe that's part of my problem. Maybe I've been too wrapped up in it."

Gina stared at him, bewildered. "But you're on the squadron commander's list. You can't just walk away when you're this close to getting a squadron."

"Sure I can. By the way, I turned down the 428th, when Spitz relinquishes command next spring."

"You did what?" Gina jumped to her feet. "Are you crazy?" She eyed him suspiciously. "Tuck, are you in some kind of trouble?"

Tuck threw his head back and laughed a low, angry kind of laugh that sent chills up her spine. "This is going to take awhile. Sit back and enjoy the ride."

Gina gulped and curled up at the end of the couch. Tuck paced again.

"It all started the day I got the phone call about Sweenedog."

For once, Gina listened without interrupting, too stunned for words.

When Tuck finished, Gina said, "I can't believe Colonel

Dennison was so stupid. Didn't the idiot think about getting caught?"

"You know what they say," Tuck quipped and paused in front of her. "All dick and no forehead."

Gina shook her head. "And Linda Garrett's no better. You know who I feel sorry for? That young sky cop. You think he'll ever trust another officer again?"

"Not around here." Tuck stopped to wind the clock.

The longer Gina pondered the situation, the angrier she got. "I gather the cover-up isn't confined to this base, since our esteemed wing commander thinks he's bullet proof. Who's the asshole protecting him?"

Tuck had his back to her. "General Honeycutt. Although I suspect the cover-up may go higher up the chain of command."

"General Honeycutt?" Gina couldn't believe it. "But he's Lois' brother."

"Yep, and he's also Dennison's boss." Tuck closed the glass cover on the clock. "I guess that tells you where the general's loyalties lie."

"I think you should contact the IG," Gina said.

"The inspector general? What? Are you kidding! The fix is in from higher headquarters. Dennison would come out looking like a Boy Scout."

"But Tuck, what they did was wrong. They deserve to be punished. And you deserve a squadron."

"Gina, it's like wrestling with pigs. It's messy and everybody gets muddy. The only ones enjoying themselves are the pigs."

Gina folded her hands in her lap. "So, what happened to the world's greatest fighter pilot? You just throw in the towel and walk away, and let him get away with it?"

"Gina, listen to me. I've had it. I want out, while I'm still ahead of my game."

At that instant, a haunted look passed over Tuck's face

and a flicker of pain returned to his eyes. Gina knew what Tuck couldn't bring himself to admit. Not even to her.

She stared at her hands. "You mean like what happened to Jeff Sweeney?"

Tuck dodged her question with a shrug. "The cover-up has merely clarified a few things for me." He gazed at something on the wall.

"What will you do about a job?" Gina took a deep breath.

"The airlines are hiring."

They stared at each other for a long time to bridge a gap that had come between them.

Finally Tuck broke the silence. "You look beautiful," he said. " We haven't had the house to ourselves in years." He waggled his brows at her.

Gina felt self-conscious and reached for the tendrils of loose hair falling around her face. "I haven't had a shower. We went straight to the mall after I got back from running."

He took a step towards her. "So."

Gina gazed at Tuck's face, at that wanton look in his warm green eyes, daring her to take him on.

She shoved her worries aside and gave the moment to Tuck.

"Come here," she whispered seductively, rising off the couch. She stepped out of her running shorts, pulled off her sports bra, and tossed them high in the air like confetti.

"You're beautiful," he said in a husky voice.

She danced toward him, bathed in the light of his love. Her full, firm breasts jiggled with each step.

"Not bad for an old lady?" she giggled.

He took her in his arms. "Best looking old lady I've ever seen." He nudged her on the lips.

Fingers trembling, Gina tussled with the zipper on his flight suit.

"I love you," he whispered. His hands explored the curves

of her back, which sent little quivers of desire into every cell of her body.

Gina motioned toward the bedroom. "Last one in bed's a rotten egg."

They ran down the hall like sex-crazed teenagers. Tuck peeled out of his uniform in hot pursuit. Gina squealed with delight when he grabbed her and threw her on the bed.

Afterwards, as they lay curled together, hot and sweaty from their love-making, Gina heard the Bronco pull into the driveway.

She sprang from the bed and dashed into the living room to retrieve her clothes.

"You're gonna get caught," Tuck laughed from the bedroom.

Gina jumped into her shorts and raced into her top when the front door swung open and the sounds of happy chatter filled the house.

Gina heaved a sigh of relief. The kids were home safe. And she finally knew what troubled Tuck.

One week later, Gina panted up the driveway after an early run. Pulling off her sweatband, she wiped her eyes and bent down to get the newspaper. Already sticky out at 6:30, she unrolled the *Bolton Daily Press* dated Monday, July 16. Her eyes widened at the headline running across the top of Page One: Beauregard's Wing Commander — Reassigned To Pentagon — Effective Immediately.

She tore through the house, found Tuck in the bathroom, shaving, and shoved the paper under his nose.

"Well I'll be damned," he grunted and continued to run the razor over his face.

Chapter Eleven

Tuck headed home from work that afternoon when a Honda Civic pulled up behind him. The driver beeped his horn and flashed his headlights, a signal for Tuck to pull over.

When Tuck spotted the driver's blue beret, he swung the 'Vette into the day-care center and waited for the off-duty sky cop to step out of his car and lope to Tuck's window.

"Sir, I have orders to Alaska," declared Sergeant Willis Chambers. "I've tried to get stationed there for years."

Tuck gave him a high five. "Congratulations. Eielson or Elmendorf?"

"Eielson, Sir!"

"I take it you like to hunt and fish?"

"Yes, Sir. I aim to bag a grizzly and catch a boat full of salmon while I'm there."

"Eielson was my last duty station before coming here," Tuck said. "You'll love it."

The sergeant stuck out his hand, turning serious. "Colonel Westerfield, despite the circumstances, it's been a pleasure to know you."

Tuck returned the handshake. "You're a good sky cop, Sergeant Chambers. Don't let the bastards get you down."

The sergeant snapped a salute, climbed back in the Honda and left.

Tuck went home to tell Gina the rest of the news. According to Bull Spitz, the Garretts had a joint assignment to Homestead Air Force Base, Florida. While Captain Linda Garrett awaited her early promotion to major, her husband, Hank, would zip around the Florida Keys in an F-16. For most military personnel, assignments never happened this fast. Especially dream assignments.

"Whistle-blowers or the threat of being court-martialed," Tuck mused to Gina when he walked in the door. "Two sure-fire ways to get a commander to give his subordinates what they want."

***

Wednesday afternoon Gina uncoiled the water hose and hauled it across the front yard. She bent down to hook it to the sprinkler and heard Wynonna talk to her dog.

"There ya go, Angel," Wynonna prattled, clutching the end of a leash. "That's a good place to potty." The Pekingese hiked up its hind leg, peed on a box hedge in front of Wynonna's house, then scurried off, dragging his mistress behind him.

"Hello," Gina chuckled.

Wynonna toddled toward her, trying to rein in Baby. "Hold on, little man." She looped the leash tighter around her hand. "Hey there," she sighed breathlessly and reeled Baby in inch by inch.

"Did Michelle ever return your bra?" Gina stood up.

Hair drooping, make-up melting, Wynonna looked like a wilted flower in the muggy heat. "There's no rush." She glanced over at Gina's house. "Is she around?"

"She's watching television."

Wynonna perked up. "Send her over."

"You got it," Gina grinned. She turned to go. "I wish you and Tuck would bury the hatchet."

Wynonna sniffed. "I'm not the one that started it."

Gina measured her words carefully. "Wynonna, how much do you know about guns?"

Wynonna stiffened. "Hardly nothing. I know a bullet's what killed Big Sandy's brother in a hunting accident."

"I'm sorry," Gina replied, "I didn't know. But that gun Tuck has, is just a pellet pistol. It might sting for a while, but that's about it."

Wynonna closed her eyes for a second, trying to digest this.

"I'm not saying what Tuck did was right," Gina added quickly. "But he's been under an enormous amount of stress. Sometimes people do things they normally wouldn't do when they're under a lot of stress."

"Is that why he's retiring? 'Cuz of stress?"

***

"Your daddy should've come to me first, instead of flying off the handle," Wynonna said, giving the can of mousse a vigorous shake. She squirted some into her hand and worked it into Michelle's short, choppy hair.

"Dad's not so bad," Michelle stared at her freshly scrubbed reflection in Wynonna's dining-room mirror. "He even went to church last Sunday."

Wynonna's mouth sprung open. "Good gravy, Angel, did the rafters fall in?"

"You sound like my mom," Michelle laughed. "She says Dad's going to hell for killing all those people in Vietnam. She says people in the military are nothing but a bunch of cold-blooded killers."

Wynonna massaged the girl's scalp in the hopes it would make her hair grow faster. "What do you think?"

"I think Dad was just doing his job."

Using her fingertips and a blow dryer, Wynonna went to work bringing order to the punk hairdo. "Until this bird's nest grows out, you'll have to camouflage it." Wynonna spoke over the hum of the blow dryer. "I suggest you get yourself a good conditioner and some hair combs. Or those elastic headbands that were all the rage in the sixties. I saw a model sportin' one on the cover of *Cosmo*."

"Mom won't let me buy magazines with sexy girls on the cover," Michelle said. "She says fashion models are pawns of the devil."

Wynonna gazed into the mirror with sympathetic eyes. "What else does your mama say?"

Michelle wiggled her blonde eyebrows, free of the heavy black pencil she'd worn into Wynonna's house. "She calls make-up devil's paint."

Wynonna threw her head back and laughed. "I don't reckon I'll be doin' much business with your mama."

She turned off the blow dryer and placed a variety of lipsticks, pencils, and other Purple Passion products on a tray for Michelle to choose.

"Here," Wynonna gestured to a row of lip pencils. "No offense, Honey, but that black junk you wear makes you look like a vampire."

Michelle picked up a rosy pink pencil and passed it to Wynonna. "I wear it to piss her off."

"Of course you do." Wynonna squatted in front of Michelle. "You're a teenager."

She applied mascara, a light dust of powder and blush, and the pink lip liner.

Baby pranced into the room and nudged against his mistress.

"Hello, Angel," Wynonna said in a sunny voice. "Mama loves you, too." The Pekingese curled up under the table and went to sleep.

Michelle peeked under the tablecloth. "You really love him, don't you?"

"I sure do. Generally, I find it a whole lot easier talkin' to animals than to people."

"Why, because they don't talk back?"

Wynonna giggled, smoothing clear lip gloss over Michelle's lips. "No, 'cuz they don't reject ya like people do."

When she finished, Wynonna pulled herself up on her three-inch mules and stepped back to examine her work. "You sure do look purty, Angel."

Michelle admired her new look in the mirror. "Mrs. Sandford, do you think I have what it takes to be a model?"

"I sure do," Wynonna chirped. "Lord knows you've got the height, bone structure, and body. The hard part's gonna be convincin' your mama."

Michelle formed her lips in a suggestive pout, as if posing for a camera. "I won't need her permission when I turn eighteen."

"Tell you what," Wynonna cut her a deal. "I'll supply you with plenty of free Purple Passion if you promise to ditch that Dracula make-up."

"Mrs. Sandford, you've got yourself a deal." Michelle reached up to shake her hand. "You're a real saint, you know that."

"Why thank you, Angel," Wynonna cast a furtive glance at Michelle's combat boots. "Now, if I can just get you into some of my little shoes..."

***

On Friday morning, July 20, a moving van and a smaller step-van pulled into the circular driveway on Bayou Way. A small army of men filed out of the smaller van and hauled sheets of cardboard boxes, ready-to-be-assembled, into the wing commander's quarters.

By mid-afternoon, most of the larger pieces of furniture were wrapped in heavy quilts and waited in the driveway. The truck driver, a grizzled-looking man in his fifties, walked around with a clipboard, jotted stuff down and stuck orange tags on every item that came out of the house, then loaded it all onto the truck.

The wing commander's wife sipped instant coffee from a Styrofoam cup and directed the movers to which items went into storage and which to Washington, D.C. Once Lois hired an attorney she would demand half of Glen's retirement and a nice fat settlement to get her started in her new life. *I'll definitely want a new wardrobe, anything that doesn't make me look like a frumpy colonel's wife and probably go back to teaching math or get a job in a bank. I like working with numbers. Numbers don't lie. I'll take a few books, small pieces of furniture that belonged to my mother, my son's baby pictures, my scrapbooks from college, the microwave, a few toiletries, and a minimum of kitchenware. An efficiency apartment won't require much. And of course I'll take the Volvo and good set of luggage.*

In turn, Glen can have the rest of the furniture, the china and crystal, the two silver punch bowls, the silver coffee service, all the items a general will need for entertaining.

She walked back through the empty house and went into the kitchen to make herself one last cup of coffee before the movers packed up the microwave.

Lois filled an old Pyrex measuring cup with water and nuked it on high. She scooped out the last granules of instant coffee with a plastic spoon when her husband walked in, wearing his dress blues and his usual air of superiority.

"Looks like you've got everything under control," he smiled tightly, his eyes swept the empty countertops and cupboards. "Sorry you got stuck handling things alone. I thought those meetings would never end." He set his keys and Geodex on the counter. "Where's the paper cups, Hon?"

"We're out."

The colonel grimaced and went to the sink.

"I'm having my things put in storage." She watched him

drink from the faucet.

"Storage?" he slurped.

The bell dinged on the microwave. Lois grabbed an old dish towel and poured water into her cup. "I'm leaving you, Glen."

The colonel whipped around and wiped his mouth with the back of his hand. Just then two movers came past, hoisted a washing machine onto a dolly. They nodded to the colonel and kept on going, careful not to bump into walls as they rounded the corner, then disappeared.

"I'm filing for divorce," she said quietly. "You'll have to handle Washington alone."

The color drained from his face.

"I'm calling Paul tonight," she informed him. "Don't worry, Dear, I'll spare him the details of your, eh, how shall I put it, your philandering."

"Who the hell have you been talking to?" he hissed. "Marsh?"

"Marsh?" Her voice cracked as she struggled to comprehend. "Dear God." She dropped her head in her hands. Her small chest caved in under the weight of a thousand lies.

"Oh God, Lois, I'm sorry," Dennison groaned. "I didn't mean to..." He stopped and closed his eyes.

Slowly, Lois raised her head. Her chest swelled like a tiny sparrow filling its lungs. "To let it slip out?" She tried to sound brave. "So, how long has that son-of-a-bitch protected you? Months? Years?"

The colonel hung his head.

"That long, huh?" Lois whispered hoarsely.

Dennison's skin turned ashy. "You're going to destroy me, aren't you?" he mumbled.

"You're pathetic," she said. "No, Glen, I wouldn't do that, because it would destroy our son. Paul thinks you and Marshall walk on water. Getting him into the Air Force Academy was easy. Making sure he graduates isn't."

She picked up her coffee to leave. "Paul has to work twice as hard as you or Marshall ever did to keep up his grades." She blew on her coffee. "Now how would that look for him? Having to explain to the other cadets that his Gods, both ring knockers, the two men he worships most in the world, have let him down, and embarrassed the academy and the entire Air Force."

"I'm sorry," he offered meekly.

The colonel walked slowly to the terraced patio, slumped down slack-jawed and awkward into a wrought iron chair.

Lois drew a deep breath. "If you'll excuse me, Glen, I have a phone call to make." She took her coffee and left.

***

"Did I catch you at a bad time?"

"Well, hey, Mrs. D," Wynonna exclaimed, recognizing Lois Dennison's voice. "I'm just sittin' here doing paperwork. Ya out of hand cream again?"

"No. Actually, I wondered if you had time to give me a make-over?"

Wynonna scurried over to the window, peeked outside, and saw a giant moving van parked across the street in front of the wing commander's house. "Well, sure I do, but won't I be in the way?"

"We'll set it up in the master bath. The movers haven't packed the vanity chair yet."

Wynonna scrambled into the dining room to refurbish her make-up kit. "Don't you worry about a thing," she chirped into the phone. "I'll have you fixed up and lookin' all purty for Washington D.C."

"I'm not going to Washington," declared the voice on the other end. "I'm headed to Mexico. Want to come?"

"Oh, you're funny, Mrs. D," Wynonna giggled.

"Bring your color wheel," the woman told her. "I want the works."

Chapter Twelve

Wednesday evening, August 1, Michelle perched on a barstool and twirled her hair. Tuck sat next to her and watched Gina rip open a bag of microwave popcorn.

Austin sat with his elbows propped up on the counter. "You're supposed to let it cool off first."

Tuck winked at Gina across the counter.

Gina made a face. A little cloud of steam puffed out before she dumped it into the bowl.

Music from *Top Gun* drifted in from the living room, where Jesse left the VCR on again. Jesse asked for chocolate milk so Gina plunked a plastic tumbler down in front of him and handed him a straw.

"Don't make a mess." She poured Austin a glass of Kool-Aid and turned to get herself something to drink.

Tuck watched his two sons plow into the warm popcorn. "Y'all act like you're starving to death." He belched up onions from Gina's meat loaf.

Despite a slight case of indigestion, Tuck raked up a handful of popcorn and settled back to enjoy it, still amazed at how quickly you could make popcorn nowadays. "Gina," he smiled, "Whatever happened to my old popcorn popper?"

Gina chuckled something under her breath and slid down on a stool next to Jesse. "You mean the one that sparked and spit at you every time you plugged it in? Why, Honey, that old popcorn popper bit the dust a long time ago. Along with the rice cooker, the electric carving knife, the fondue pot, and that god-awful fuzzy bedspread that looked like a dead buffalo."

Austin and Jesse laughed.

Tuck felt a bit wounded. "I always liked that bedspread."

"Sorry," Gina sipped a glass of rosé.

Scratching his chin, Tuck recalled other missing household items. "Whatever happened to that ceramic monkey I use to have? You know, the one playing golf. And those big throw pillows with the leopard print?"

Gina blinked. "The ones that smelled like cat pee? Gone by the way of the dead buffalo, and the rest of that herd of white elephants I inherited from your first wi—" Gina stopped and blushed instantly.

"Loose lips sink ships," Tuck eyed her wine glass. Gina dropped her gaze at something on her lap.

Twirling her hair, Michelle seemed oblivious to their conversation. "Guess what?" she said as if in a daydream.

Tuck turned sideways. For a fleeting moment, Michelle looked just like her mother. Without thinking, Tuck called her Sally.

Michelle blinked.

"Jesus," Tuck laughed, "I guess I had a brain fart."

"Fart-fart-fart," Jesse jabbered then slurped chocolate milk through his straw.

"I've been called worse," chuckled Michelle.

Tuck hoped Gina wouldn't be upset. *At least I never slipped and called her Sally.*

Gina nibbled on a piece of popcorn, watching him like a hawk.

Tuck turned to Michelle. "I'm sorry, Sweetheart. What

were you saying?"

Michelle's finger stopped midtwirl. "A lifeguard at the base pool offered me a hundred bucks to..."

"To do what?" Tuck barked, cutting her off.

Michelle jumped. Her finger twirled faster and faster around the wispy tendrils of hair below her earlobes. "To pose naked," she squeaked.

"Naked," Tuck said aghast.

"He only wants to snap a few photos to send to *Playboy*," she added. "He says he can make us both a bundle of money and jump-start my modeling career."

"Over my dead body," Tuck growled, then crammed a fistful of popcorn in his mouth. "Trust me, Sweetheart. That lifeguard wants to do more than jump-start your career."

Jesse looked up from the end of his straw.

"Uncle Bo gots pictures of naked womens on his walls," he announced, "but we're not allowed to go in there, are we, Mommy?"

"That's right," Gina pet the top of his head.

Tuck took a deep breath.

"How old is this creep?" He shoveled more popcorn in his mouth.

"Don't talk with your mouth full," Austin cut in, flicked his tongue back and forth in the double-wide gap where his front teeth use to be. "Mom says it's bad manners."

Tuck's head spun around. "You look like a lizard when you do that."

"Don't encourage him," Gina said.

Tuck turned to Michelle. "Well, let's have it. How old is this guy?"

Michelle dangled her long, brown legs as if on a boat dock gazing at her reflection in the water. "Old enough to wear dog tags," she volunteered.

"Damn airman," Tuck snorted, picked a piece of popcorn out of his teeth with his fingernail.

Michelle stared out the window into the night.

Gina's eyes narrowed on Michelle. "I hope it's not that nice young airman from the chapel?" Her voice rose an octave as she went to wipe a chocolate mustache off Jesse's upper lip. "Didn't Chaplain Lawrence say he was a photographer at the base photo lab?"

"Better not be," Tuck jumped in, "or he'll be singing soprano the rest of his life."

Michelle slid around. "Why do you two always think the worst? And no, it's not the same guy."

"I liked you better when you looked like Beetlejuice," Tuck grumbled and stretched back to yawn. "At least I didn't have to worry about every swingin'-dick from here to eternity trying to get in your pants."

The air conditioner kicked on, sending a blast of cold air into the room.

"What's a swinging-dick?" Austin reached for more popcorn.

Michelle buried her face in her hands.

Tuck glanced helplessly at Gina.

"Well now," Gina's eyes rolled thoughtfully. "A swinging dick, my dear, is what your father was before he married me."

Tuck laughed and walked to the refrigerator.

"I don't get it," Austin said.

Tuck popped open a beer. "You will when you're older." He took a sip, then pulled out a gallon of milk.

Austin plopped his face between his hands. "That's what you always say."

"Is this milk any good?" Tuck bent to sniff it.

Gina swiveled around and frowned. "I hope so." She picked up Jesse's empty glass and waved it in front of her nose. "And don't drink out of the carton."

Tuck put the milk away and turned with a grin. "I'm gonna need an Instant Breakfast in the morning. I'm back in the saddle."

Slowly, Gina set the glass down. "You're back on flying status?"

"Spitz called me into his office this afternoon."

Jesse jumped off his barstool, knocked it over, and went flying down the hallway toward the bedrooms.

Gina bent over to pick up the stool. "Guess your back's on the mend."

Tuck tipped his beer high in the air, clinking it against an imaginary glass. "Guess so." He offered Gina a sip.

She guzzled the beer like water. Wiped her mouth, then she passed it back. "What time do you fly?"

"O-dark-thirty. I'm gonna finish this beer and head to bed."

A flicker of fear shot through Gina's eyes.

Tuck set his beer down. He held out his arms, and Gina buried her face in his chest. "When you were grounded I didn't worry about you."

Tuck lifted her chin. "I'm the world's greatest fighter pilot. Remember? I'm invincible. I'm Superman."

A weak smile crossed Gina's face.

"I hope you remember how to fly," Austin stared cross-eyed at something on the tip of his nose.

Glancing at his son, Tuck caressed the small of Gina's back. "Hey, it's like a riding a bike. Once you learn how, you never forget." He nibbled on Gina's ear, causing her to giggle.

Michelle kept a close eye on the two of them. "Does this mean you won't get to retire after all?"

Tuck looked at his daughter. "No, Honey. It means I won't have to fly a desk the remainder of my tour."

Gina pulled away from him and started to clean up the kitchen.

Michelle brushed salt off her hands and unfolded herself from the stool. "You'd look pretty stupid flying a desk around all day," she went off to start the boys' bath water.

Tuck threw his beer can away when Jesse zoomed into

the kitchen wearing Michelle's combat boots, Tuck's old flight cap, turquoise sunglasses, and the fake-leather bomber jacket Jesse got last Christmas.

"I'm a fighter pirate." Jesse spread his arms out like wings. His voice bubbled over with laughter like little bells tinkling in the wind. Despite the lump in his throat, Tuck grinned.

Jesse adjusted his plastic sunglasses and gave Tuck a greasy thumbs up. "Do some of that pirate ship, Mav," he jabbered impishly, mimicking a line out of *Top Gun*.

Tuck stuck his thumb in the air and sent Jesse off into the living room to watch the end of his favorite movie.

"Oh brother." Austin spit an unpopped kernel across the kitchen like a sunflower seed. "Jesse's gonna wear out that video."

***

Gina curled up in Tuck's chair, reading the latest squadron wives' newsletter, when Michelle shuffled into the living room, half asleep around two in the morning. She had on one of Gina's ratty old nightshirts and a pair of bunny slippers from Wynonna.

Everyone else was asleep.

"Have you heard of a little country called Kuwait?" Michelle yawned and rubbed her eyes. "It's somewhere in the Middle East."

Gina looked up. "No, it doesn't ring a bell. Why?"

Michelle plopped down on the couch and grabbed a pillow. "I heard on the radio that it was invaded by Iraq."

Gina yawned, turned a page. "It's probably some Arab squabble. I wouldn't worry about it."

Michelle cuddled up with her pillow. "Mind if I lay here awhile? I can't sleep."

"Me, neither," Gina reached up to turn the lamp down a notch. "There's an afghan over there if you get chilled."

***

Two weeks later, Tuck walked in the door for lunch. He stood at the edge of the kitchen, holding his briefcase and flight cap. He'd spent the morning attending one briefing after another, all pertaining to the crisis in the Persian Gulf. Gina put the broom and dustpan away. "I don't think it's fair that you have to go."

"Life's a bitch." He stashed his briefcase on the floor and went to wash his hands. "And that's if we go. For all we know, this thing could end tomorrow."

"Don't count on it." Gina dragged a barstool to the stove.

The house smelled of furniture polish and glass cleaner. Ever since the base was put on alert, Gina went on a cleaning binge. Each morning she ran, then spent the rest of the day cleaning house. Sometimes Tuck could hear her clean out closets, scrub the toilets, haul out garbage at all hours of the night.

Drying his hands, Tuck watched her climb on the stool and move things around inside the dinky cupboard above the stove. Tuck didn't even know what was up there.

"You're going." Gina pulled down a grimy Thermos and shook it. Tuck could hear the rattle of broken glass. Gina tossed the Thermos into the trash and shut the doors. "I can see it in your eyes. I can feel it whenever I give you a hug." She headed for the refrigerator with the trash can in tow. "You're already starting to pull away from me and the kids. It's like part of you has already left."

Tuck couldn't help it. He had to get his mind right so he could focus on the mission, in case they were tasked to go. No mental baggage allowed in the cockpit.

He crossed his arms, leaned against the sink, and watched Gina tackle the refrigerator. First she pitched jars of pickle juice, crusty mustard, jellies that turned to sugar. The sound of breaking glass filled the room. Next came half-empty bottles of salad dressing, a hunk of moldy cheese, a stalk of limp celery.

Then she pulled out a small Corning Ware dish, hidden behind a bag of sugar. Wrinkling her nose, she peeked under the plastic lid, then snapped it shut. "Yuk." She tossed the whole thing in the trash.

Tuck frowned. "What did you do that for? That was a perfectly good dish."

Gina closed the refrigerator and headed for the sink. "I think it was spaghetti and meatballs in its previous life. I'm not taking any chances."

Tuck shrugged and moved out of her way. He went to the refrigerator and rooted around in the meat drawer for something to eat. He pulled out a package of hot dogs and walked to the bread box.

He heard Gina's knees pop when she went to bend down. "You've already been in one war." She rummaged in the cabinet under the kitchen sink. "It's somebody else's turn to go."

Tuck twisted open the bread. "Who said anything about war. We don't even have orders yet."

Gina stood up. "This base wouldn't jump through the hoops if something wasn't up."

Tuck slapped a cold hot dog between a plain piece of bread and folded it up like a bun. "I'm headed to the base legal office to update my will." He pinched together both ends of the crust. "It's just part of the alert procedures. Nothing to worry about." He grabbed a Coke and leaned against the counter to wolf down his lunch. "The house looks nice, Babe."

Gina popped on a pair of rubber gloves and sprinkled Ajax over both sides of the sink. "Thanks." She ignored his remark about the will. "I'll enroll Michelle in school this afternoon." She doused her scouring pad under the faucet. "Jo-Ellen wants to introduce her to the principal and some of the staff."

Tuck bit into his hot dog. "Have her transcripts arrived yet?"

Gina turned and gave him a pensive look. "They came yesterday. She's barely passing."

Tuck stopped chewing. He thought immediately of his younger brother Bo, and the struggles Bo endured in elementary school before the folks finally admitted that maybe mainstream education wasn't the route to take for someone like Bo. By the time the folks got him enrolled in Special Ed, Bo already lost any self-confidence he might have had as a toddler.

Tuck resumed eating and kept his thoughts to himself.

Gina scrubbed the sink again. "Jo-Ellen's offered to tutor Michelle for free."

Tuck picked up his Coke. "What does Michelle have to say about her grades?" Taking a drink, he told himself, *Surely this beautiful daughter of mine can't possibly have a defective brain housed inside that pretty head. Isn't one Bo in the family enough?*

He set his soda down.

Gina stopped scrubbing for a second. "I haven't talked to her yet. But if you ask me, her grades are a symptom of having to put up with Sally."

Tuck stared at his half-eaten hot dog. "Tell Jo-Ellen we'll pay her. Westerfields don't sponge off their friends."

He crammed down the rest of his lunch when the phone rang.

Gina was up to her elbows in scouring powder and hot water.

Tuck took a chug of Coke and reached for the phone.

After hearing the stiff, formal voice at the other end, Tuck eyed Gina closely. "Hello, Raymond. Everything okay?"

Gina's whole body went rigid. Her head swung around.

"I'm not here," she mouthed. She attacked the sink, as if determined to get out some stubborn old stain.

Tuck psyched himself up for the phone call. He wondered whether all grown daughters could dismiss their fathers so easily with a flash of the eye, a sudden turn of the head. He took the phone into the living room. At the sliding glass door, he caught a glimpse of his own daughter sunbathing on the patio. She had a beach towel tented over her head, reading a magazine. Austin and Jesse squatted in the sandbox, spooned sand into old pots and pans and pretended to cook.

Turning away, he made the usual excuses for Gina. She's at a friend's house, on a walk with the kids, at the commissary buying groceries.

Gina rinsed out the sink when Tuck walked in to hang up the phone.

"Don't tell me," she said.

Tuck grabbed his keys and bent to give her a kiss.

"Dad's company's going to bend the rules and offer combat insurance. Double indemnity if you get your head blown off. Triple if they can't find it."

"He just called to check on you. See if we knew any more." Tuck put on his flight cap. "By the way, he said to tell you he loves you."

Gina hung her head. "Oh," her voice cracked when she spoke, "I'll call him later, after Jesse goes down for a nap."

Tuck picked up his briefcase. "I'll stop by the supply shop after work to pick up my chemical warfare suit. If I'm not home by supper, y'all eat without me."

Gina nodded and turned away, as if she wanted to be alone.

***

Bull Spitz paraded back and forth in front of the TV, brandishing the remote control like a laser gun. Every few seconds he stopped, aimed the clicker at the TV, and zapped it to another channel.

In the kitchen, Becky Spitz cleared her husband's TV tray and dumped the rest of his BLT down the disposal, followed by a pan of bacon drippings.

"Turn that damn thing off!" her husband barked from the other room.

Becky dropped ice cubes into the dark hole while the gears were still grinding, followed by a squirt of lemon juice, to cover the odor. "Yes, Sir, your highness." Flipping off the switch, she brushed toast crumbs into the sink, then swept into the living room in a ruffled apron with blue geese cross-stitched on each pocket. "Aren't we in a cranky mood." She gathered the morning newspapers splayed across the couch.

Bull jerked his head around. "Beck, I can't hear the news."

She bent over the coffee table, rubbed at something dark and gooey with the back of her apron. "Bull, why don't you go back to work? I'm sure they won't start the war without you." She picked up the candy dish, plucked through it for empty wrappers.

Bull whirled around, his forehead bulged. "Becky, for crying out loud, this is the real thing. It's not some piddly-ass exercise where we fly to the range and back."

Becky stood up to look at him. "Don't I know that," she bristled and reached around to untie her apron. "Ever since the base was put on alert, you've stomped around here like Attila the Hun." She pulled the apron stiffly over her head, folded it even though it was dirty. "Why do you think I've rushed around to get organized, in case I have to throw a party at the last minute."

"You and your parties," he grumbled. "You gonna have everybody dress up as insects again?"

"Oh hush." She fluffed up the sofa pillows. "This is going to be a bon voyage party. I'd shirk my responsibilities if I didn't send your men off properly."

"Beck, we're not going on a cruise. Besides, you just threw a farewell party for Colonel Dennison."

"So."

"I'm going back to work." He switched off the TV and stalked toward the door. He stopped long enough to check his hair in the entry-way mirror and pluck his flight cap off the hall tree. "What time do you pick the boys up from day camp?"

"After I make a pass through the commissary. Why?"

"I thought you might like to join me for a drink at the club."

"I can't." She swept her folded apron over the top of an end table. "The kids offered to help me work on decorations tonight." She paused. "The guys can dress up as Arab sheiks and the girls can come as belly dancers. What do you think?"

Spitz called halfway out the door. "You gonna call it *Arabian Knights* and have the wives show up with fake rubies in their navels?"

Becky clutched her apron to her breast. "Why Bull, what a marvelous idea. Why didn't I think of that."

Once the Saab was out of the driveway, Becky hurried into the kitchen, clawed her way through a stack of tea towels and pulled out a white envelope postmarked Acapulco.

The upper left-hand corner of the envelope was blank. She stared at it a moment, feeling a strange mixture of apprehension and excitement. The rest of the day's mail sat unopened on the counter where she'd dropped it when Bull popped in unexpectedly for lunch.

Even with Bull out of the house, she couldn't take any chances. She took the envelope and locked herself in the master bathroom. She slit open the envelope and read and prayed she had done the right thing.

Dear Becky,

Take my advice. A stiff drink and a little sun does wonders for a woman's soul. Not to mention a jazzy skirt, a peasant blouse, and a touch of Purple Passion, that magical potion Wynonna Sandford markets.

In my rush to leave Beauregard, I failed to thank you for clueing me in about Glen. Telling me took a great deal of courage. Please don't fret or lose any sleep over this. I've lost enough for the both of us. Besides, what you told me only confirmed my suspicions.

I'm sorry I had to bow out at the last minute, but I trust the going-away party you threw for Glen was on a grand scale. I heard the decorations were clever as always: stars made out of tinfoil, hanging from the ceiling of the O-Club. Rather fitting for a man about to make brigadier general.

Tomorrow I go para-sailing. I know it's hard to believe, but I haven't had on a suit or pantyhose since the change of command. And forgive me, my dear, but I hope I never have to attend another social gathering where I have to wiggle my hips in a grass skirt, sit around in a white sheet, or two-step my way across the dance floor in some raunchy ranch duds I've borrowed from the Base Thrift Shop. Not that the Hawaiian Luau, Toga Party, and Western Night weren't fun, Dear.

Bye now,
Lois

Becky stood over the toilet and shred up the letter.

She jerked the handle, then watched the incriminating evidence swirl away. "The woman never looked good in a grass skirt anyway," she mumbled. The toilet gurgled back at her.

It hummed for a while, then stopped. Becky dropped the lid, cracked open the frosted window above the toilet, and peeked out. She was greeted by brilliant sunlight and a strip of green grass. The house next door was jammed so close to hers she rarely opened her blinds, much less a window, unless her oddball neighbor, Vic Rollings, was TDY.

But there was always an exception. Like the night the exhaust fan in her bathroom went on the fritz, and Becky had no choice but to crack open the window and air the place out after her shower.

Becky's mouth tightened at the memory. Bull had gone next door to tell Vic Rollings to turn down his music. Rollings hosted one of his UFO parties, and by the time Becky stepped out of the shower and cracked open the window, the music stopped.

While the party guests peered through telescopes on Rollings' roof, Bull and Vic Rollings stood on the side of the house, talking. Becky overheard their discussion as she toweled off.

The two men retold the story over and over: How they drove up behind the barn and caught Colonel Dennison and Linda Garrett trying to put their clothes back on. And Tuck Westerfield, poor drunken son-of-a-gun, stunk like a brewery with his hands up, while a nervous young sky cop tried to figure out exactly who to point his gun at.

Convinced she had done the right thing, Becky snapped the window in place and bustled out of the bathroom. She had a party to plan.

***

August 23, Gina opened the sliding glass door and poked her head outside. "Michelle, it's your mom." With a pit in her stomach, Gina held out the phone.

Michelle looked up from the swing set, where she watched the boys roll Hotwheels down the slide into the grass.

"Is anything wrong?" she asked, coming toward Gina. She had on a pair of cut-offs and the top to her new bikini.

Gina cupped her hand over the mouthpiece. "Your mom's leaving South America on the next plane. She wants you back in Galveston in a couple of days. She said some mumbo jumbo about the Holy Land and the crisis in the Middle East."

Michelle rolled her eyes and took the phone.

Gina was about to leave, to give Michelle more privacy, when Michelle grabbed her by the arm and motioned for her to stay. "What should I tell her?" she hissed and held the mouthpiece against her bare midriff. "I'm not going."

Gina glanced at the swing set. "Tell her the boys need you desperately in the next few weeks."

Michelle took a deep breath and put the phone to her ear.

In a matter of seconds, Gina could tell Michelle lost the battle. Sally didn't buy it. Michelle's voice grew quiet, submissive, timid. Her shoulders drooped and her bare feet turned inward, making her look pigeon-toed.

Gina's heart sank. Then she got angry.

"Hand me that phone."

Michelle thrust it at her and ran to the edge of the patio. "Tell her to go fuck herself," she cried and burst into tears.

Gina covered the mouthpiece. "What did Sally say?"

Michelle looked at her with puffy, red eyes, eyes full of suppressed rage and a good dose of guilt. "Same old thing," Michelle headed for the door. "She told me to get ready."

"Ready for what?" Gina frowned, holding the phone.

"The Apocalypse," Michelle snorted, slinking into the house.

***

Two days later Tuck and Gina left the house to put Michelle on an airplane. In two more days, Tuck was bound for the Middle East.

Tuck knew Gina had done everything in her power to persuade Sally to let Michelle stay. And once again, Sally won. She had custody.

No one said much on the drive out. Gina stared out the window and clutched her purse. Michelle sat in the back seat with her arms draped maternally around each of the boys.

Tuck drove with his jaw clenched. Seven weeks ago, he and Gina had driven down this same stretch of road to pick up the kid because she had no place else to go.

At the airport, Tuck parked the truck and they went inside. While Michelle marched up to the counter to get her boarding pass, Gina touched Tuck on the shoulder. "You polished her boots."

"It's the least I could do," Tuck shrugged.

Last night, when Gina and Michelle said goodbye to Wynonna, Tuck slipped into Michelle's room. What he found was a room caught in an identity crisis. Although Jesse bunked in Austin's room, his little-boy bedspread still covered the twin bed, and on the walls, pictures of farm animals and airplanes competed with posters of rock stars and fashion models.

At the foot of the bed, like defiant old soldiers, stood his daughter's scuffed-up boots, laced up with bright pink shoestrings for the trip home. Shaking his head in dismay, Tuck went to get his polish and rags.

With her boarding pass in hand, Michelle strode across the terminal toward them in a pink mini-skirt and denim vest, oblivious to the commotion she caused, particularly among the men. One guy almost walked into a door, he stared so hard.

Outside, the air felt heavy, still, as if the world held its breath. Michelle kicked at a rock on the sidewalk and waited for her flight to be called. Tuck fiddled with his change. Gina patted the boys on the back.

"Sorry I won't be here tonight to baby-sit." Michelle glanced at Gina. "Aren't you suppose to dress up as a belly dancer?"

Gina chuckled and threw Tuck a wicked glance. "Your dad's gonna don a turban and strut into the O'Club like he's Ahab the Arab."

Tuck stood with his arms crossed, gazed at them behind his sunglasses. "I think I'll pass."

The airport intercom crackled to life. "Flight 197 to Galveston is ready for boarding."

Gina wrapped Michelle up in her arms and murmured something in her ear. Then she turned away and grabbed a Kleenex out of her purse.

By now, Austin sniffled and Jesse cried, "Stay here, Sissy."

Michelle bent down, hugged them both, then turned to face Tuck.

Tuck cleared his throat and tried to sound stoic. "Your mother won't recognize you." He embraced her.

"When did you stop loving her?" Michelle whispered in a desperate voice.

Tuck held his daughter at arms length. "You mean your mother? Not until I met Gina."

Michelle nodded thoughtfully, as if her father finally cleared something up that churned inside of her for a long time.

"You think we'll go to war?" she stalled for time.

"Who knows. That's for the politicians to decide."

"Saddam Hussein said his people will eat any American fighter pilot shot down over Iraq. You think he's bluffing?"

"He's bluffing," Tuck said with a little more bravado than necessary.

"Are you afraid?"

Tuck fixed her with an insidious grin. "I don't get paid to be afraid, Sweetheart."

"Final boarding call for flight 197..."

Tuck motioned for her to go. "You better skedaddle."

She gave him a peck on the cheek, glanced one last time at Gina and the boys, then with her boom box and suitcase slipped through the gate and headed out to the airplane. At the bottom of the rollaway stairway, she turned one last time and blew them all a kiss, then started up the stairs, right boot-left boot-right boot-left boot and smacked each steel step with purpose.

As Tuck watched her mount the stairs, he had a revelation. Maybe, just maybe, those battered combat boots represented the very thing Sally hated the most, the military. Not to mention the ex-husband who served in it.

At this most inappropriate time, while the pilots revved up the engines, and Gina and Austin sobbed and Jesse screamed, "Sissy come back. Sissy come back," Tuck threw his head back and laughed like a wild man.

"What's so funny?" Gina blew her nose.

"It's those damned boots," he laughed. He was still chuckling long after the turboprop rolled down the runway and lifted into the blue.

Cupping his hands around his mouth, Tuck hollered at the tiny black dot on the horizon: "When the shit gets too deep for you, Sweetheart, put on your boots."

***

When Tuck got home from the airport, a letter postmarked Colorado Springs waited for him.

Dear Tuck,

I'm here in Colorado Springs visiting my son, Paul. He's in his third year at the Air Force Academy and plans to make a career out of the Air Force. Paul is unaware of his father's recent activities. I want him to finish up here and get through pilot school before I burst his bubble.

The reason I'm writing is this: I simply cannot make a clean break if I have to keep going to bed every night with you on my conscience. By now I'm sure you've figured out that my brother, Marshall Honeycutt, is the man responsible for protecting Glen. They met at the Academy, long before I fit in the picture.

In order to make sense of my brother's misplaced sense of loyalty when it comes to Glen, I think it's important to understand the dynamics of their relationship. After Glen graduated from the Academy, he joined up with Marsh in Vietnam, where they flew F-4s in the same squadron.

They flew together the day Glen got shot down. I don't know all the details, but I suspect Marsh is to blame. Neither will talk about it. After Glen's release from prison, Marsh took Glen under his wing and has bent over backwards to help him. As you can see, Glen has used this to further his career.

I understand that most of Beauregard will deploy to the

Middle East. I assume that includes you. My prayers are with you and your family. Give my regards to Gina. She always stood apart from the crowd, and for that I admired her.

Bye now,
Lois Dennison

Chapter Thirteen

Monday, 0600 August 27. Behind the podium, Lieutenant Colonel Bull Spitz's fervent speech came to an abrupt halt.

"Ten hut."

Inside the briefing room of the 428th Tactical Fighter Squadron, forty pilots jumped to their feet and snapped to attention.

Colonel Rich Maple, Beauregard's new wing commander, a tall, big-boned man with an easy gait and wide-set eyes, strode into the room, followed by Colonel Marv Blevins, the vice wing commander, and Colonel Ron Abbott, the DO, deputy of operations.

Colonel Maple advanced toward the podium.

Bull Spitz stepped quickly aside as Colonel Maple turned and addressed the room. "At ease, gentlemen." He broke into a lopsided grin.

A collective sigh went out over the room as the pilots took to their seats.

Tuck sat in the third row back, directly behind Tony. Wheaties and Killer were up front, next to a guy who kept biting his nails.

After Colonel Maple said a few words, he turned the briefing over to Colonel Abbott, who went over last minute details and answered questions.

The pilots had been given their flight plans in advance. Their destination: a classified air base in Saudi Arabia. The wing commander would lead the deployment. During his absence from Beauregard, Colonel Blevins would stay behind as the acting wing commander.

As far as Tuck knew, neither the DO nor the vice wing commander had a clue about what transpired over the summer. Both men had been away during most of that time, either on lengthy TDYs or on extended leaves.

From what Tuck gleaned from Bull Spitz, the new wing commander, brought in from Alconbury, a base in Great Britain, didn't have a clue either.

With his chin propped on the tips of his fingers, Tuck put the whole thing out of his mind and focused on the mission ahead. Compared to a possible confrontation in the Middle East, Dennison's escapade seemed like child's play.

Near the end of the briefing, Chaplain John Lawrence walked quietly to the podium, opened his Bible, and read:

"To the one we are the smell of death; to the other the fragrance of life."

He snapped the book shut.

"If the time should come to liberate Kuwait," he said sternly, looking out over the room, "remember that verse."

Tuck had never heard the chaplain speak with such conviction in his voice.

Lifting his palm to offer the benediction, the chaplain passed his hand slowly through the air and paused at different intervals along the way as if to stop and touch each man on the forehead before he continued. "Now go with God," he said. "You have my blessings."

The chaplain gathered his Bible and left. He had another squadron to bless, and bags to pack. That night, the chaplain flew out on a C-5, along with hundreds of other troops from Beauregard.

After the briefing, the pilots grabbed their flight gear and left. Tuck, the last pilot out of the building, spotted Tony and Wheaties waiting for him at the door.

The sun crept up over the horizon and looked like a giant orange beach ball, heralding the end of summer as it floated above a dim line of trees.

Up ahead, Bull Spitz and the rest of his posse of freedom fighters, as Spitz called them, made their way across the ramp as a bevy of local reporters and photographers scurried nearby to get the story of the year.

Tuck clamped his jaw, stared straight ahead while he and his two companions marched by in front of the press.

"Zip-a-dee-doo-dah, gentlemen," he quipped out of the side of his mouth, "We're on Candid Camera."

"And a mighty-high-ho-fuckin'-silver," Tony bantered back and grinned at the cameras.

"Nothin's too good for our boys in blue, hey Drivers?" Wheaties joked then held up a peace sign.

In front of them, loaded down with long-range fuel tanks for the trip across the Atlantic, sat their beloved warthogs.

Unlike the pointy-nosed jets with their cosmic computers and navigation systems, the A-10 was strictly a stick-'n'-rudder aircraft. No autopilot. No magic. A demanding mistress, yes, and if a pilot couldn't handle her, she did a lot more than complain about it. She'd kill him in a heartbeat.

"It's show time, boys," Tuck declared. The three of them split off in different directions, each heading to his own airplane.

Tuck took a deep breath, filled his lungs with the heady scent of jet fuel.

He was pumped.

PART II

Letters

Chapter Fourteen

Gina browsed through the September issue of *Family Circle*, mesmerized by the picture of a succulent pork roast.

Tuck loved pork roast, though she seldom fixed it. No telling what kind of mystery meat he got in those rations the military called MREs, Meals Ready to Eat.

She ripped the picture out of the magazine, along with the recipe, and stuck it up on the refrigerator. "For Tuck's first night home," she scribbled at the top of the page.

It was a Monday, the tenth day of September. Tuck had been gone fifteen days. Outside, the twelve o'clock whistle went off that reminded her to pick up Jesse from the base preschool.

Every day at noon, the command post tested its emergency warning system by blowing its siren. To the men with empty stomachs on the flight-line, on duty since sunrise, the daily siren sounded more like a lunch bell.

The short burst reached its peak and wound down by the time Gina grabbed her purse and keys and cracked open the door.

"Ohmygosh!" She halted in the middle of the doorway and propped the storm door open with her hip.

Her black, middle-aged mailman flashed a cheerful, gold-toothed grin. With a flourish, he waved an envelope in Gina's face.

"Special delivery," he announced, then winked at Gina and passed her an envelope postmarked Air Force Postal Services. Tuck's Social Security number and return APO address were printed neatly in the upper left-hand corner.

Gina brushed her fingertips over Tuck's handwriting, as if she could feel him through the ink. She looked up and smiled. "It's the first time I've heard from my husband since he left."

The mailman beamed as if he already knew this. "Yes Ma'am," he hoisted his mail satchel stuffed full of letters from Saudi a little higher on his shoulder. "Ya'know, Miz Westerfield, I b'lieve I'm startin' to feel like Santy Claus."

Tues. 31 Aug.
(The middle of nowhere)

Dear Gina,

We got here today at 0745 your time, 1545 Saudi time. My ass is dragging. It was sporty flying over the Atlantic at nighttime, heavy thunderstorms and lightning. During one of the refuelings, a couple of guys lost sight of the tanker and declared lost wing man, but other than that, we all made it here in one piece. A tad saddle-sore, but glad to see dry land.

This place sucks. It's a huge, unfinished airport terminal with two gigantic runways in the middle of the Arabian Desert. Not much to look at but miles of caliche-colored sand, blue sky, and a brutally hot sun. Reminds me of something straight out of the movie *Road Warrior*. Any second now, I expect Mel Gibson to drive up.

My hooch isn't fancy. It's a twelve-by-eight room at one end of a tiny trailer—but it beats living in a tent like the enlisted folks. I share a bathroom with a guy on the other end. Wheaties and Tony bunk four to a room in some trailers nearby. Spitz is in the trailer next door to me. Hey, at least we don't have to share the toilet.

No word on when we get to fly. For now, I'll work at the command post, help set up the operations side of the house. Except for the runways, and a bunch of trailers left behind by construction workers. We build a base from scratch. God, what I'd give for a cold beer.

Kiss the boys for me,
Love,
Tuck

P.S. Any news from Michelle? Here's a letter for the boys.

Dear Austin and Jesse,

Are you guys behaving yourselves? Austin, does that pack of girls from kindergarten still chase you on the playground? (A guy should be so lucky.) And Jesse, how's Daddy's big boy? Can you say your ABCs? Tell Sissy hi for me when you talk to her. Kiss Mom for me.

I miss you guys,
Love, Daddy

August 28
Beauregard AFB

Dear Tuck,

I miss you and you've only been gone two days. The boys had trouble sleeping last night, so I put them in bed with me. Austin cried himself to sleep. Jesse wet the bed. (Sorry, your side.) First accident he's had in weeks. The first thing out of his mouth after he woke up this morning was: "My daddy's coming home today." Austin and I didn't have the heart to correct him.

I'm sure by the time you get this letter, Wheaties will have announced to the whole squadron, and half the Middle East, that Sylvia's pregnant. Due the end of March. Except for a terrible case of morning sickness, she's ecstatic. Jo-Ellen and Krystal already plan a baby shower.

Becky Spitz stopped by the house today to drop off an invitation for a Mexican fiesta she's hosting for the wives. Honest to God, the woman wore a peasant blouse and a sombrero. She looked like a cross between Carmen Miranda and the Frito Bandito. Then she took Vera Maples to lunch. That's the kind of crap I'd do if you were a squadron commander—shuttle the new wing commander's wife around and throw kitschy theme parties. La Cucaracha! Let's party.

There's a town meeting at the base theater tomorrow night. Colonel Blevins will update us on the operation and show slides. It's a shame Jo-Ellen and I can't sit together, but she has to support her maintenance wives. Krystal has to work late, and Sylvia offered to baby-sit the boys. She can get to a toilet a lot faster at our house than if she tried to feel her way through a dark, crowded theater. (I hope Jesse doesn't poop in his pants while she's there.)

FYI: Since you can't have any booze over there, I gave up my nightly glass of wine. How's that for solidarity?

Gina

Sun. 2 Sept.
(The middle of nowhere)

Hi Babe,

I got baptized today at one of our weekly Morale Meetings. Chaplain Lawrence did the honors. We can't call it church because it might offend the Saudis. Armed guards are posted outside in case some nosy Saudi wanders in during The Lord's Prayer or a Hail Mary. The chaplains run several services each Sunday for both Protestants and Catholics, flip-flopped back and forth to accommodate all the shift changes. The Jews meet on Saturday for synagogue.

After the service, I told Wheaties I had a hunch my grandmother rolled over in her grave the second the holy water hit my head. All those years, she must've thought her two grandsons got baptized in New Orleans on one of those rare weekends when we didn't visit. How else can you explain the fact that she didn't personally slam-dunk us in the immersion tank of the Rose Glen Baptist Church?

As for the climate here, I thought Arizona was hot or Louisiana. But nothing compares to this kind of heat, except a blast furnace. Once the sun goes down, it cools off real fast.

Love, Tuck

3 Sept.

Dear Austin and Jesse,

Today is Labor Day. Sorry I'm not there to make homemade ice cream. When I get back, we'll buy an electric ice-cream maker from the BX. Tell your mom to plant geraniums in the old one and have the handle bronzed. Have fun during the day and sweet dreams at night.

Love, Daddy

5 Sept.
(The middle of nowhere)

Hi Babe,

I got your letter today. It doesn't bother me if the boys sleep with you once in a while, as long as they don't make it a habit. I got to fly today for the first time. Tell the boys this place is one big sandbox. Visibility is poor. It's hard to tell where the sand stops and the sky begins on the horizon. Without air conditioning in the cockpit, we'd fry. Under these conditions, our canopies act like a hothouse. (Think twice about those skylights you want in your dream house.)

We flew out to the west about 150 miles, and saw nothing but sand dunes, a few camels and goats, a family or two of Bedouins—nomads—and some U.S. Army tanks. We rocked our wings a couple of times at the grunts. They love hog drivers—it's our job to save their butts if a war starts.

When we're not flying, we help the civil engineers fill sand bags and build tent city, the area where we house the enlisted folks. Big Sandy acts like he runs the show. I'm sure he gets a big kick seeing a bunch of cocky fighter pilots sweat their asses off doing hard labor.

Love, Tuck

P.S. Wheaties struts around like a barnyard rooster and bugs Tony, "You next, Bro?"

September 6, 1990
3702 Bayou Way

Dear Dad, (Mom is helping me write this)

We tied a yellow ribbon on the pecan tree next to the driveway. Mrs. Sandford tied a yellow ribbon in Baby's hair and a big yellow bow on top of her purple car and the other car Mom calls *The Yellow Submarine*. If it was black, it would look like The Batmobile.

Miss Jo-Ellen put yellow paper all over her door at school, but some mean kids ripped it up. Mom says they got in big trouble. Some other kids bought Miss Jo-Ellen flowers and more paper and helped her fix her door.

I am doing good in school. My teacher's name is Mrs. Miller. She looks like Miss Krystal. There's a girl in my class named Allison and her mom works on airplanes. She had to go to Saudi Arabia too. Allison's grandmother came all the way from Maine to stay with her because Allison doesn't have a dad. Well, my hand is getting tired and Mom is getting tired of spelling. I ask Jesus every night to keep you safe.

Love,
Austin

P.S. Jesse wants to know if you've ridden a camel or shot any bad guys. Also, can we get a dog? Mrs. Sandford told us where we can get one for free. She said if nobody takes them home, they will die.

***

Sept. 8, '90
Mrs. Wynonna Sandford
3704 Bayou Way
Beauregard AFB, LA

Ms. Katherine Tuttle (CEO)
Purple Passion Beauty Inc.
Fort Worth, Texas

Dear Ms. Tuttle,

Hey there. Remember me, your top-selling beauty specialist for 1989? Have I got a great idea for you, and a way for Purple Passion to support the troops in Saudi Arabia and boost our sales at the same time.

Wouldn't it be patriotic if Purple Passion sent over a big shipment of complimentary products for our men and women who sacrifice in that gritty desert while the rest of us sit back here powdered and perfumed in our air-conditioned homes and cars.

If our military folks love our products, and you know they will, chances are they'll want to order more when they get back to the states. I hope to hear from you real soon.
By the way, some of my most passionate customers are Air Force wives, not to mention quite a few fellas in the Navy.

Yours truly,
Mrs. Wynonna Sandford

***

Sept. 10

Dear Tuck,

Jesse stepped in a pile of fire ants today after I picked him up from preschool. I rushed him to the emergency room on base.

Doc says he'll live, but Jesse's still pissed about the shot they gave him. The whole time, he screamed, "My Daddy's coming home today."

It's all my fault. I was on the phone with your mother when I heard his bloodcurdling scream. Both of his legs and scrotum are covered in ant bites. After the boys get a bath, I'm headed out to put down more poison.

Surviving on the home front,

Love,

Gina

September 11, 1990
3702 Bayou Way

Dear Dad,

Mom declared war on the fire ants last night. She said if we're not careful, fire ants will take over the world. She put on a pair of your old flight boots and a face mask and went outside and shook a big bag of poison all over the yard.

Jesse looks like he has chicken pox. I feel sorry for him, but you told us to stay away from those ant beds and Jesse didn't listen.

Your son, Austin

***

Purple Passion Inc. Ft. Worth, TX
From the desk of Katherine Tuttle, CEO

Sept. 15, 1990
Dear Mrs. Sandford,

What a brilliant idea. Of course Purple Passion will support Operation Desert Shield by sending free products.

Marsha Burns, our national director of marketing, will handle the arrangements.

Below is a list of items we thought might be useful to our men and women deployed to the Middle East:

—our new SPC 30 sunscreen (chemical free)
—lip balm (for both men and women)
—hand and body lotion (scented and unscented)
—our total hair care program (including our new line of men's products)
—cleansing gels and soaps
—our new and improved waterproof mascara
—our expanded line of deodorants

We hope this meets with your approval. If we've left anything off, let us know. Keep up the good work.

Warmest regards,
Katherine Tuttle

***

23 Sept.

Dear Jesse,

I hope you're feeling better. Daddy should have put down more poison before he left. Tell Mom it's not her fault, and thanks for the bag of lollipops.

I know you guys want a dog, but now is not a good time. We don't have a fence, and it would be cruel to keep a dog chained up all the time. Tell your brother a house dog is out of the question. Trust me boys, I had a dog when I was growing up, and they're more trouble than they're worth. They poop, bark, bite, get fleas, shed, smell, get sick, and die. How about goldfish?

Love, Daddy

October 1, 1990
3702 Bayou Way
Beauregard AFB, LA

Dear Dad,

Grandma Westerfield called to check on us. I told her you got baptized in Saudi Arabia because she forgot to get you baptized when you were a baby.

Grandpa Loyd called and I told him we didn't need any more insurance. Mom says I can't answer the phone anymore. She says I'm too smart for my own good. But Mrs. Miller says it pays to be smart. She says I'm the smartest boy in the class.

Love, Austin

P.S. I wrote a story about Jesse and the fire ants. I said Jesse's balls got all swollen. Mrs. Miller laughed when she handed me back my paper, but she said she didn't think the girls would like to hear about that.

October 31

Happy Halloween!

I took the boys to the Pumpkin Patch off base the other day and let them pick out pumpkins. We spent all afternoon carving jack-o-lanterns.

Jo-Ellen took the boys trick-or-treating tonight, so I could stay here and pass out candy. Austin went as the Lone Ranger and Jesse as Tonto. Jesse threw a hissy because I wouldn't let him go barefoot. Jo-Ellen dressed up as a witch—her alter ego after teaching all day. Sylvia and Krystal hid out at the base theater. (*Steel Magnolias* played for the umpteenth time.) Obviously, Sylvia's feeling better. She sure was sick as a dog that first trimester.

Speaking of dogs, trust me, Tuck—to use one of your favorite phrases—the boys aren't taking "No" for an answer. And fish? Come on Tuck, like your oldest son said, "Since when can you teach a fish how to play fetch?" Wynonna says the animal shelter has lots of animals for adoption. By the way, you never told me you had a dog growing up.

Woof Woof, ya old sober-sides,
Gina

P.S. I had lunch with Jo-Ellen in the teacher's lounge today. You wouldn't believe the streets of Bolton. Yellow ribbons everywhere, from trees to store fronts to people's mailboxes. On the way to town, I even saw a couple of scarecrows decked out in yellow.

1 Nov.
(The Middle of Nowhere)

Dear Gina,

Just finished writing a letter to my folks. Everything here's about the same. We're all bored as hell, even with the all the flying and other duties. Wheaties and Tony run a couple of miles each day around the compound to relieve stress. They keep bugging me to join them. I tell them God only gave me so many heartbeats and I'm trying to conserve them. We've all lost weight over here. Nobody pigs out on MREs.

For the most part, I'm doing okay. But sometimes this place gets to me. Even when things went to shit with Dennison, I could still walk away. I can't do that here. Stay safe.

Love, Tuck

Nov. 13

Dear Tuck,

A cold snap hit here a few days ago. The temperature got down to thirty-five last night. The trees look naked and gray, except for the live oaks and pine.

Right now, I'm in bed buried under a pile of covers to stay warm. Outside, the wind whipped up the leaves something fierce. I can hear them scuttle down the street. If you step outside, the air has that cold, smoky smell that I love. I wish we had a fireplace.

I read some trashy romance novels Jo-Ellen loaned me. Granted, it's not great literature, but it sure makes you horny. I miss you.

Love, Gina

November 30
New Orleans

Dear Tucker,

Daddy said you sounded plain tuckered out in your last letter. Bo got confused at your daddy's choice of words. I explained to your brother that what Daddy meant by tuckered out was you sounded tired. Bo said: "Well that's a fine thing to name a son. A word that means tired." Imagine, your brother coming up with something like that. And they told me Bo was slow.

We didn't do anything special for Thanksgiving. Bo pitched a fit when I told him we were having it catered, but your daddy stepped in and told him how lucky he wouldn't be sitting over there in Saudi Arabia under a tent eating Army chow with President Bush. Albertson's deli prepared a nice bird with all the trimmings, but their pecan pie wasn't as good as mine and their dressing rather bland without mushrooms and oysters.

We called Gina today. We're sending her some early Christmas money. I don't have the energy to shop anymore. Your daddy came into the room and told me CNN's reported there's talk of an air strike sometime after New Year's, if Saddam doesn't pull his troops out of Kuwait. Now Tucker, no matter what happens, I want you to know that Gina and the boys are always welcome here. And that Austin, he's a smart little boy for his age.

Love, Mama

Christmas Day

Dear Gina, Austin, and Jesse,

HoHoHo from the land without snow. Thanks for the flannel shirts and the reading lamp. I've got it clipped to a nail over my cot. And the extra batteries and new underwear.

Austin, did Santa bring you your two front teeth? You know the song: *All I want for Christmas is my two front teeth.* And Jesse, how's my big boy? Did Santa bring you that red pedal airplane you wanted?

Gina, for your eyes only. Did Santa bring you any trashy novels to snuggle up with while I'm gone? You know, the ones that make you horny. If he didn't, go get some, because when I get home, I want you to climb me like a tree. I miss you so much it hurts. I love you, never forget that, no matter what happens.

Tuck

P.S. I meant to tell you, Purple Passion has invaded the desert.

Chapter Fifteen

January 16.

"Come Mommy come," Jesse shouted from the living room. "Come and see."

Gina browned a pound of ground beef on the stove. The overhead fan sucked up steam.

"Come and see what?" she hollered back and ripped open a box of Hamburger Helper.

"Fireworks!" Jesse shrieked in a voice so loud and shrill it cut through the sizzle of meat and the drone of the fan.

Gina froze, held her breath. "What fireworks?" She turned off the fan.

"The fireworks on TV."

A tingle of fear crept up her spine. *Today is the deadline for Saddam to pull out of Kuwait.* She stirred the crumbled meat, elbow macaroni, and tomato-red powder lumped together in the skillet. While part of her body shut down—the part that tried to digest inconceivable news—the other part continued to function. Her hands had a mind of their own. She tossed the empty box in the trash, went to the sink for water— exactly three and three/fourths cups—then back to the stove to pour, stir, cover, simmer and set the timer.

One of the boys cranked up the volume on the TV. She knew the sounds coming from the next room. Popping noises—the chatter of anti-aircraft machine guns—the low murmurs of TV reporters dispensed doom and gloom.

Gina felt strange moving about in her own kitchen, as if part of her wasn't there. Her hands, feet and legs propelled her in motion, while the rest of her floated out into a cushion of air, trying to breath. She could hide there forever.

"Mom, you-better-get-in-here," Austin hollered. "They're shooting guns."

Gina slid the skillet off the heat and turned off the stove. She could no longer ignore the boys' fervent calls, or the dry, metallic taste in her mouth.

Her heart hammered. She sucked in her breath and stepped around the corner of the kitchen, into the line of fire. A spray of florescent green explosions shot skyward across the top of the television screen, like the video games the boys played at Chuck E. Cheese's. Gina's knees buckled at the sight of the boys, huddled shoulder-to-shoulder in front of the TV. She stared at the flashes of light rip open the canvas of night over downtown Baghdad. She knew that somewhere up there in the dark, at the top of the TV screen, the coalition fighters whizzed past in the veil of night and rained their own brand of death and destruction down on the city.

"It sounds like popcorn," Austin observed and turned to look at her.

Her teeth chattered, but Gina hugged herself and inched closer to the TV, as if some invisible vacuum sucked her forward against her will.

"Mom, are you cold?"

"No. I'm fine." She shivered from fear, and didn't want the boys to know. She had on a long, shaker-knit sweater, a warm pair of leggings, and thick cotton socks. The thermostat, set at a toasty 72 degrees, staved off the cold, damp air that nipped at the windows.

"Daddy's not up there, Mom. Trust me." Austin seemed to read her mind.

"How do you know that?" Gina glanced at Austin.

"Because the A-10s aren't fast enough for the surprise attack," he explained. "They need planes that fly high and fast, like Stealth fighters and F-16s and F-llls."

Gina stared at her son, amazed at his knowledge, at the level of insight he displayed. Tuck would be pleased.

"They're saving the Warthogs for the good stuff," Austin declared and puffed up his chest like General H. Norman Schwarzkopf himself. "They're there to help the grunts and blow up tanks."

"Bang bang," Jesse roared and fired make-believe pistols at the TV screen. "Kill the bad guys. Kill the bad guys," he chirped to the tune of Wagner's *Cry of the Valkries.*

Even if Tuck wasn't involved in the first airstrike, Gina knew his time would come. The danger wasn't just over the skies of Baghdad, either.

Iraq supposedly had Scud missiles—armed with chemical warheads—aimed straight at Saudi Arabia. Maybe even straight at Tuck's compound.

Gina jumped when the phone rang. She hesitated, then hustled into the kitchen. In the time it took her to grab the cordless, scurry into the other room, and drop into Tuck's chair, the slurred speech and sloppy, self-indulgent tears bombarded her.

Gina cringed, then braced herself.

"I tried to tell him," Tuck's mother rasped in her slow, drunken drawl, her voice scratchy from too many cigarettes. "I warned him those warplanes were dangerous."

Gina shut her eyes as her mother-in-law launched into another rampage. Tucker had a mind of his own. Tucker never listened to a word she said. Tucker with his wild and foolish dreams that could get him killed.

Gina felt desperate. She could hang up, take the phone

off the hook, and say later that they'd been disconnected.

She heard a scuffle, the sound of the dropped phone, then the kind, craggy voice of her father-in-law breaking in to rescue her.

"Sorry about that, Gina. The old woman's hit the sauce since Christmas. She's done got Tucker dead and buried just like she did in Vietnam." He reflected on something. "If you ask me, this live coverage makes it worse. Y'all doing all right?"

Gina felt a catch in her throat. "How's Bo?" she blurted, desperate to take her mind off the TV and her mother-in-law's worries.

"Bo?" Her father-in-law seemed vaguely surprised that she inquired about Tuck's brother at a time like this. "Oh, he's doing all right, I guess—for Bo. He's wired for sound, you know," he continued with a chuckle. "Between his hearing aids and those headphones he wears twenty-four hours a day, and a new satellite dish we got him for Christmas, he keeps us posted on who's running for what, which team is in the playoffs, which general is top dog for the day, and which Playmate of the Month is featured on the *Playboy* channel." He chuckled again. "Which by the way your mama's not too keen on..."

Tuck's dad did that lately. Referred to Gina's mother-in-law as your mama.

Gina loved her father-in-law. She felt sorry for him hobbling around in his walker. The polio that ravaged half his body when he was a young man left him crippled in old age. Yet he still had a sense of humor.

"Bo has already worked on his wish list for next Christmas," Tuck's dad said. "He asked for a big-screen TV and one of those Ataris." Tuck's dad acted like every aging parent had a forty-year-old son, gray at the temples, still living at home.

"You mean a Nintendo?" Gina corrected him.

"Is that what it's called? I can't keep up with all these new fangled gadgets."

Gina stared at the TV, numb.

Her father-in-law's voice grew thick all of a sudden. "Here, Bo has something he wants to tell you."

It pained her the way Bo looked at her sometimes, cross-eyed behind his Coke-bottle glasses, his slightly stooped body still small, like an adolescent boy, hunched over as if afraid to stand up straight to face the world.

"G-Gina, it's me, B-Bo. D-don't you worry ab-about my brother Tu-Tucker now," he stumbled over every other word.

Gina swallowed. Sometimes it hurt to listen to him struggle.

When Bo was little, kids used to call him "shit-for-brains" and "Bo-the-Retardo." Tuck told Gina he spent half his childhood defending his brother from bullies, and the other half resenting him.

"G-Gina...?" In that tentative voice again, afraid he wasn't worthy of addressing other people. "Tu-Tucker, he's like a b-bird. Our gr-grandma always said Tu-Tucker knew how to fly, ev-even before he was b-born."

Gina's throat constricted. "I know, Tuck told me."

"G-Gina..? He's a real sm-smart man, my brother, like a wi-wise old owl." He paused. "Pe-people don't shoot down owls, do they, Gina?"

"I hope not." She choked on her own words. *For a guy with shit for brains, Bo is pretty damn smart.*

After they clicked off, Gina gathered herself in a ball and burrowed in for the long haul.

Moments later she punched in Jo-Ellen's number.

Jo-Ellen picked up on the first ring. "Gina, is that you?"

"I'm scared shitless," Gina confessed, staring into the TV.

"I know. So am I."

"Mommy," Jesse hung his head, "I pooped in my pants again."

"Oh God," Gina shut her eyes. "I can't take anymore."

A few minutes later, a maroon New Yorker zoomed up the driveway and screeched to a halt. Jo-Ellen Hawkins barged into the house like she owned the joint.

"Pull yourself together!" she ordered. "Don't you dare wimp out on me now."

***

Jan. 18, 1991

Dear Tuck,

When I was young, I used to visit Aunt Cora at her funny-smelling farmhouse on the outskirts of town. The house had a long front porch that stretched from one end of the house to the other. I liked to sit on the porch and take in the view of the air base. The highway, railroad tracks, and a chain-link fence separated us, the farmhouse people, from them, the military. The front porch had the best seat in the house. Unfortunately, the rattlesnakes thought so, too.

Nevertheless, I spent hours camped out in a rusted metal lawn chair and clung to an old garden hoe in case Mr. No-Shoulders decided to slither up through the honeysuckle and join me while I waited for the jets. The boom of the F-100 Super Sabers' afterburners caused my aunt's teacups to rattle, the windows shake, and the life blood pump through my veins.

Then a big, dark jet would shoot straight out of nowhere over the top of the farmhouse and scream like a mighty bird of prey. Its steel belly so close I knew if I jumped high enough off the porch, I could reach up and touch it before it landed on the other side of the fence and rolled down the shimmery black runway and out of my life.

For one brief, shining moment, nothing else mattered. Then I went back to look for snakes. You know Tuck, in all those years I never did see one, dead or alive, near the premises. Yet I heard all about them in the stories Aunt Cora would tell anybody who listened—like how she nearly died wrestling the devil's serpent with the blade of her hoe. In the world of snake kills, the woman was an ace.

Looking back, I realize I could have stayed inside and peered out the picture window in Aunt Cora's living room instead of venture out into the danger zone of the porch, where things could bite. But I had to see the jets and feel the thunder they created in the atmosphere. Their arrivals and departures made my daydreams take flight. The jets so close, yet so far away, in a world I wasn't privy to. A world my dad once moved in and out of as an airman, but left before I was born. A world I didn't think dangerous, until I married you.

Last night, when they showed footage of an A-10 from Myrtle Beach shot down over Iraq, the war came crashing into my living room. After all these years, Tuck, one of those goddamn snakes has finally crept beyond its boundaries.

It pisses me off. I'm sitting here 7,000 miles away on the edge of your chair, armed with nothing but this lousy remote control. So I've decided to take matters into my own hands. As soon as I stick this letter in the mail, I'm headed out to the shed to fetch the garden hoe and tie a yellow ribbon on it.

Love You,
Gina

***

Jan. 18
3704 Bayou Way

Hey Angel,

This smelly bum stops by the shelter the other morning while I'm feeding the cats. He looks worse than Sarge, a skinny German shepherd I brought home last week. (More on that later.) Anyway, he says he's down on his luck and maybe a pet would cheer him up.

So I steer him over to a black cat I call Bad Luck. I'm

crossing my fingers because Bad Luck has, as they say at the shelter, worn out his welcome, and they're putting him to sleep the next day. The man looks at me like I'm crazy and says, "Lady, don't you reckon a black cat with a name like Bad Luck is the last thing I need right now?"

I tell him politely, "No offense, Mister, but from the looks of things, they couldn't get any worse. Why don't you give some of that reverse psychology a try—you know, where you do the opposite and hope for the best." He screws up his scruffy face like he's thinking real hard, and says, "Tell you what, lady. If you throw in some cat food and an old blanket, I'd be happy to take Bad Luck off your hands." I made him promise me that if things didn't work out, he could bring Bad Luck back, no questions asked. What the world doesn't need is another poor alley cat.

Now about Sarge. I know we agreed a long time ago that I wouldn't feel sorry for every four-legged critter that comes through the door. Sarge is different. There's something about his eyes—so lonely and pitiful and wise. I know he can't stay here forever, but I couldn't walk away and let them put him to sleep.

The Westerfield boys have taken a shine to him. The minute I get home from work, they're at my door to see him. They want a dog but Colonel Grumpy won't let them. Even if they were to get a dog, you-know-who would probably use it for target practice.

Love and kisses,
Wynonna, Baby, and Sarge

***

Jan. 20

Hello Trumpeter,

The doctor moved my due date up two weeks to mid-

March. Gina's offered to be my labor coach in case you're not back in time. She says having a baby is a piece of cake. This is a woman who ran outdoors in Alaska at thirty below and opted for natural childbirth.

Here's a side shot of me at the last coffee. Becky Spitz is having it blown up into a poster to hang in the squadron rec room, along with dozens of other pictures she's making into a giant collage titled "Meanwhile — Back on the Homefront." She already planned a homecoming reception, complete with a giant warthog cake made by a lady in Natchitoches. Krystal says it's the same lady who made the armadillo cake for the wedding scene in *Steel Magnolias.*

By the way, your Uncle Matthew called while I unloaded the groceries. He told me he'd been out all morning "strafing rice paddies full of little gomer bastards" in his A-1 Skyraider. I take it that meant he'd been out in his Piper Pawnee spraying for boll weevils. Either he's suffering from delusions, or he's got a bad case of post-traumatic stress syndrome. It's gotten worse since you left. You might want to drop him a line.

Love,
Sylvia

1 Feb.
Nude beach in Saudi

Hey Driver,

Just kidding. There's not a naked lady in sight. Saudi women come with bags over their heads, down to their toes. Seriously, thanks for checking on Sylvia. I've had a lot of time to think over here. Not much else to do when I don't fly, eat, or sleep. I guess that's how it was for you in Southeast Asia. Anyway, I thought a lot about that afternoon after Daddy's funeral, when you told me to go get in the airplane, we'd take her up for a spin.

Do you remember me throwing a fit? I screamed I'd

never crawl inside another airplane as long as I lived? Mama, if I recall, stood there all tightlipped like she had a corncob up her butt, and let you carry me off to the hanger. Then we took off, you in the front seat, me in the back, with the wind in our faces and Mama shrank behind us. Before I knew it we buzzed the rooftops and highways like the most natural thing in the world.

Thanks, Uncle Matthew. When this war's over, I want you to come down to Louisiana for a visit after the baby's born.

Live to Drive,
Wheaties

***

Feb. 3

Dear Tuck,

Austin scared the crap out of me after school today. He watched TV while I ran to the bathroom. Jesse took a nap. When I came out of the bathroom Austin vanished with the TV on and he hadn't touched his snack. I checked everywhere, under the beds, the closets, the shed, the truck, the neighbors, before I called the sky cops. I thought someone had come in and snatched him right out from under me.

While the sky cops had a canine cop working the banks of the bayou, Wynonna Sandford and a sickly old German shepherd she rescued from the animal shelter went down to the base playground and found Austin huddled up inside the A-10. A logical place, don't you think? The sky cops were embarrassed, but they're undermanned since most of them are in Saudi.

By the way, the dog died.
Gina

P.S. The mailman asked me why I have a garden hoe with a yellow ribbon propped outside the front door. I told him it was to ward off evil spirits. He thinks I'm joking.

Feb. 4, 1991
3702 Bayou Way
Beauregard AFB

Dear Dad,

Mom took me to a shrink today at the base hospital. The doctor asked me why I ran away yesterday. I told her I got scared, because another A-10 got shot down and the bad guys came and captured the pilot. I thought it was you. The doctor asked me if I thought I could really fly that airplane at the playground all the way to Saudi Arabia. I told her, "No dummy, that A-10 doesn't have an engine. It's fake." Mom took me back to school and she said I don't have to see the shrink anymore. What bugs me is why a nice dog like Sarge had to die. I guess he didn't have any life insurance because Mrs. Sandford didn't get rich.

Your son,
Austin

Feb. 6
Galveston, TX

Hi Dad,

You should let the boys get a dog. A dog would help them take their minds off the war. They may be young, but little kids understand a lot more than adults give them credit for. I'll even throw in a flea collar, if that helps.

I'm a working girl now, baby-sitting three hellions after school each day. Mom only agreed because they're Pastor's kids. And Pastor and his wife can't find anyone else who can handle the three little turds. Whenever they get out of line, all I have to do is threaten them with an hour's worth of Bible memorization. That shuts them up.

Mrs. Sandford sent me another care package yesterday, and Mom threw a fit. First she bitched about the way I used to look. Now she bitches that I look too desirable. She says I'll never get into heaven if I don't clean up my act. I told her I already had.

Oh yeah, I made the honor roll. Mom says scholars are the scum of the earth. Sometimes I think she'd be happier if I'd been born deaf, dumb, and blind. Or not at all.

Love,
Michelle

***

Feb. 7

Hi Angel,

Sarge earned his stripes in the hero department. Unfortunately, he died in the line of duty. When Austin Westerfield came up missing the other day, and the sky cops couldn't get their act together, I had Gina bring me one of the boy's dirty shirts and held it up to Sarge's nose. His ears perked up when I told him the shirt belonged to the little boy next door.

Sarge took one sniff of that shirt and pulled himself up off the floor, his skin sagged so badly I thought I'd have to pin it up with a hair clip. Once he picked up the scent, he galloped down the street, panted louder than a locomotive and pulled me with him. I nearly broke my neck trying to keep up.

By the time we reached the playground, Sarge breathed so hard I could hear his bones rattle. He was like a horse drawn to water, charged out across the playground to that airplane you built last year. When he stuck his nose down in the hole and licked at something, I knew we found our boy. Austin crouched down inside, shivered and cried that he had to rescue his daddy. "From what?" I asked, and he said, "The bad guys."

By the time the sky cops showed up and gave us a ride home, Sarge was on his last leg. Bless his heart, he stumbled out of the patrol car and looked up at me with those sad brown eyes of his, then rolled over and collapsed right there in the

Westerfield's driveway, before any of us had a chance to say "Thank You."

I think the Air Force ought to give Sarge the Medal of Honor or at least a Purple Heart.

Love,
Wynonna and Baby

13 Feb.

Dear Wynonna,

I'm sorry about Sarge. He must have been one hell of a dog in his day. But if anyone deserves a medal, you do. Colonel Spitz nailed me after a meeting the other day and asked if y'all shipped more hair spray. That guy uses so much hair spray, if a fly were to land on his head, it'd get stuck there and die. But who's going tell a brute like that to tone it down. Anyway, I said I didn't know, but I told him I'd be happy to place a special order for him. He balked when he realized he would have to pay for it this time. Another pilot looking for a free ride.

I bumped into Westerfield on my way home from work tonight. He was civilized for a change, and for some god-awful reason, I told him I washed out of pilot school. I told him how I tried to take Clem's place after he tripped on that damned barbed-wire fence and blew his head off. He asked me if Clem had the safety off, and I said, "Yessir, and I always thought cadets at the Air Force Academy were smarter than that."

Westerfield laughed his ass off and said, "Don't count on it." I'm not sure what compelled me to stop by his hooch in the first place, but the both of us talked more tonight than we have the two years we've been neighbors.

War's a funny thing, Schnookums. It can bring out the best and the worst in people.

I Love Ya Girl,
Big Sandy

P.S. I'm running low on deodorant.

***

13 Feb.

Hi Babe,

Still dodging bullets and flying daily missions over Iraq and Kuwait. Big Sandy stopped by my hooch tonight while I sat on the stoop. He was bored like the rest of us.

I told him how grateful we were for Wynonna's help when Austin ran away. Did you know Wynonna lost a baby years ago, and that she was abandoned by her own mother at birth? Left her at a filling station, next door to the pound. Big Sandy told me some other stuff too, but I'd rather discuss it when I get home. It helps explain why he has a grudge against pilots.

Love,
Tuck

Feb. 13
3702 Bayou Way
Beauregard AFB

Dear Dad,

I saw a picture of a homeless man and his black cat in the newspaper today. The man and his cat found a hundred-dollar bill in a trash can while they looked for food. Guess what the cat's name is? Bad Luck. The newspaper said: "With a cat named that, who needs a good-luck charm?" The man said he used to be a soldier a long time ago, when people spit on soldiers for going to war. I hope nobody spits on you. That would be gross.

I asked Mom if we could get a cat. She said don't even think about it. Besides, me and Jesse still want a dog. Mom said to ask you, if you can't teach an old dog new tricks, can you at least get it to heel?

Your son,
Austin

Chapter Sixteen

Feb. 13
Galveston, TX

Hi Dad,

Pastor freaks me out. He tells us to sell our worldly possessions and follow him to the mountain. He hasn't said which mountain. Or why we need to bring all our money. He says if there's a ground war, all the Bible prophecies will be fulfilled and the world will come to an end. I guess Pastor thinks that if he's up on a mountain, he can be the first person on earth to flag down Jesus at the Second Coming and welcome him back. Of course Mom believes all of this nonsense. She thinks Pastor walks on water. I told her if Pastor was so perfect, then how come he fathered three little rotten terrorists. Write me back.

Love,
Michelle

Feb. 15

Dear Tuck,

Your parents called last night. Your dad sounded fine, but your mom worries so much she's a basket case. After she told me about the roses your dad gave her for Valentine's Day, she started crying.

Mom sent the boys a box of chocolates. Jesse ate too many and got sick. He threw up all over my lap. Luckily, your chair is Scotchguarded. While I ran this morning, my period started. I'm talking major flood here, then my right knee decided to give out a mile from the house. Try limping home on one good leg with Aunt Flow springing a leak in your running shorts. That was the longest mile I ever ran in my life!

Love,
Gina

19 Feb.

Dear Michelle,

I just got your letter and, quite frankly, I'm alarmed. Doesn't this Pastor have a first or last name? I have about as much faith in some of these so-called ministers as I do in faith healers, psychics, and Ouija boards.

If your mom decides to follow this guy, get hold of Gina immediately. I repeat, do not go with your mother. She's obviously been brainwashed.

Love,
Dad

19 Feb.
(Middle of nowhere)

Dear Gina,

That church Sally's hooked up with sounds like a doomsday cult. Either this Pastor is a scam artist or another Jim Jones. I've instructed Michelle to get in touch with you in case Sally does anything stupid. Please call Michelle and check on her. I guess she's too young to remember what happened in Guyana.

I'm flying two and three times a day. Flying surges is a lot like driving a race car. You land the plane, pull up to the pits, engines running, refuel, rearm, guzzle a few drops of water, then you're off, shooting back down the runway for another go at the war.

Everybody wants to get this thing over with so we can come home. Kiss the boys for me.

I love you,
Tuck

22 Feb.
(Middle of nowhere)

Hi Babe,

Bull Spitz has earned the nickname Pedro, because he won't fly north of the border. The border between Saudi Arabia and Iraq, that is.

Buzz Hawkins says Spitz brings his plane back to maintenance all the time with bogus write-ups and emergencies. Yet Buzz's crew chiefs can never find anything wrong. If Spitz takes off on a mission and no one on the ground shoots back, he drops his bombs on his assigned targets, then flies home, business as usual. But if there's even a hint of enemy response, he high-tails it back to base with a sick jet, leaving everyone else to finish the job.

All the 428th pilots have lost respect for him, not to mention the crew chiefs. Buzz is disgusted, but he pretty much has his hands tied. Does he risk his own career and file a formal complaint, or does he go along with Spitz's little charade? It's a pretty stiff accusation, calling someone a coward. Especially when that someone is a commander who is supposed to be a leader, and you're just one of the worker bees. Sounds familiar, doesn't it?

Love,
Tuck

P.S. Any word from Michelle? Get a dog for the boys if you must. Just don't let them get too attached. In case something happens. There's rumors the ground war may kick off any day.

***

25 February. Persian Gulf, early morning, the second day of the ground war, the sun peeped over the horizon, then quickly vanished behind a bank of low clouds. Tuck and Wheaties slipped through the mist and headed north toward southern Kuwait. Their mission: to strike Iraqi forces that opposed U.S. Marines as they moved north in their drive to liberate Kuwait City.

"The weather's dog shit," Tuck said on the radio as he popped in and out of the clouds.

"Roger, One," Wheaties fired back and trailed closely behind him.

Every time a hole opened in the clouds, Tuck glanced out the side of his canopy at the ground and caught a glimpse of the U.S. Army roll toward Kuwait City. From an altitude of 15,000 feet, the battalions looked like toy tanks and trucks and little green, plastic Army men crawl along on the desert floor.

If it were only that easy, Tuck thought when his plane slipped into another pocket of fog. He turned his eyes to his gauges to guide him through the soupy sky. *The only thing plastic about war is the body bags they ship dead soldiers home in.*

A short time later, at a prearranged rendezvous point, they made radio contact with their FAC, a forward air controller in an OA-10.

Their target, the FAC pilot explained, was a column of fifteen to twenty Iraqi tanks and trucks fleeing northeast toward Kuwait City, along the highway just east of Ahmed Al Jaber Air Base.

After their FAC finished giving them their coordinates, they dropped below the clouds, and moved in for the kill.

"They're sitting ducks." Tuck eyed the designated line of tanks and trucks that sped across the desert. Though it rained the day before, they kicked up dust.

At this point in the war, Tuck wasn't too concerned about

anti-aircraft fire, although he kept a constant vigil for possible SAM sites. Most Iraqis seemed more concerned with running away than in fighting.

"Keep it above 5,000 feet until we're right over their tails," Tuck radioed Wheaties.

"Roger, One," Wheaties radioed back.

They rolled in over the column and dropped their ordnance and strafed the hell out of the column.

Fifteen minutes later, most of the vehicles on the ground turned to smoking hulks.

As the FAC gave them their mission results, Tuck spotted a lone enemy tank speed away from its burning comrades.

Tuck was at bingo fuel, but he had enough ordnance to finish off the tank and get home before his fuel ran out.

Tuck jammed his mike and called the FAC. "There's a bad guy trying to get away. I'm in from the west."

"You're cleared hot, One," the FAC signaled the go-ahead.

"I'm right behind you," Wheaties broke in, ready to back him up.

Tuck had one Maverick missile left, and he didn't want to waste it.

Adrenaline pumped through his body. Fearless, he had to get that tank, before it got one of the friendlies.

He rolled his plane into a dive and went after the tank. But when he went to lock his Maverick onto the target, the missile broke lock. He tried again, and again it broke lock.

"Watch your altitude, One," the FAC warned. "You're getting too low."

"Roger," replied Tuck, as he tried one last time to lock onto his target.

"Look out," Wheaties cried.

Tuck had no time to jink—to dodge whatever it was that was racing up to kill him.

A giant Ka-boom slammed into Tuck's jet and turned his

world inside out.

The impact was so violent it jolted his feet off the rudder pedals. The airplane shook like it was coming apart. The instrument panel lit up like a score board gone haywire; red and yellow lights flashed on and off as the cockpit filled with smoke.

So this is what it's like, huh Sweenedog, when things go to shit.

For a split second, Tuck had the oddest sensation of floating in space, suspended in a quiet cocoon of peace, and in the next second gripped with an unholy terror that seized his body and howled to his brain "punch out."

But he didn't. Not yet.

Instinct and experience took over. From what he could tell, he was still in one piece, wrapped inside the titanium bathtub that made the A-10 so survivable. *Maybe that's all that's left,* he thought in a detached sort of way. *Just me and the cockpit, whizzing through space, before gravity has its way and we drop to the earth like a two-ton brick.*

Tuck jammed his mike and called for Wheaties.

"Two, I've been hit," he kept his voice calm. "The cockpit's full of smoke and I can't see. I think I lost an engine."

"Roll right," came the swift and sturdy command from his young wingman.

For the first time in Tuck's career, a second lieutenant was in charge of keeping him alive.

He followed Wheaties' instructions and rolled right. A dose of adrenaline surged through Tuck's body and sent him the jolt of electricity he needed to survive.

"Okay, Boss," Wheaties said, "you're level. Pull up easy."

The smoke cleared, and Tuck could see the nose of his airplane coming up. As he climbed, gained altitude, Wheaties joined up on his wing.

"Well, Two, how bad does she look?" Tuck glanced over

his shoulder. He trailed smoke and leaked fuel like a sieve. The plane's controls felt stiff and hard to handle—worse than a stubborn old clunker in the days before power steering.

A long silence ensued before Wheaties replied, "Like a train wreck, Sir."

Tuck flew home on one engine, his right wing hung by a thread, the ejection handle a breath away, just in case.

Once Tuck landed the crippled jet, he didn't waste any time to shut it down. With fluids dripping all over the runway, he popped opened the canopy and bailed over the side. He scrambled down the ladder two rungs at a time in case the whole thing blew up.

He never looked backed. He kept moving, walked quickly away from the airplane as a swarm of emergency vehicles screamed across the tarmac, and encircled the area like vultures around a fresh carcass.

A Land Rover tore across the taxiway and screeched to a halt in front of him. Tuck yanked off his helmet and stared at the driver.

Bull Spitz sat behind the wheel and jawed on a piece of gum in that slow, deliberate way of his, like he and the gum in cahoots with each other plotted something sinister against Tuck. He leaned out the window and rested his elbow on the ledge.

"Well I'll be damned, if it isn't our Chief of Safety," Bull Spitz snorted, then reached up to scratch his sunburned nose. "Didn't you see the SAM coming?"

Tuck cocked his head to the side, commented on the squadron commander's blistered face instead of the insult. "Jee-zus, Bull. Did you forget to wear sunscreen?"

Spitz closed his eyes, shook his head, as if Tuck was his own personal cross to bear. "Always the joker, aren't you Tuck. Can't ever be serious about anything." He let out a sigh. "You just broke a perfectly good airplane, and now I suppose you'll get a medal for it."

Tuck laughed and stepped up to the Land Rover.

"Hey, Bull. This isn't a ladies' tea sip, you know. There's a war going on." Tuck smiled like a schoolboy.

Spitz flinched, drawing back. "You imply something, Westerfield?" he muttered, jutted his jaw defensively.

Tuck leaned closer and got a whiff of Spitz's hot, fruity breath.

"Don't you ever chew anything besides Juicy Fruit?"

Spitz quit listening. He gazed at something over Tuck's shoulder.

"You're lucky you got out alive," Spitz said at last, a hint of admiration and jealousy in his voice.

Tuck could hear the bustle of activity going on behind him. Crew chiefs scrambled around the crippled jet. Firemen hosed spilled jet fuel off the ground.

Tuck shrugged like he was use to missiles blowing him out of the sky.

Spitz scratched his nose again. He wouldn't look Tuck in the eye. "I'll meet you back at the squadron."

Tuck felt drained as he stood on the side of the runway and watched the Land Rover drive away. Putting up a front robbed him of his last ounce of energy. The burst of adrenaline that pumped through his veins over the battlefield like a rocket booster launched him into survival mode, had long since burned up. He ran on empty.

His arms and legs felt weak and useless, virtually nonexistent, jettisoned from his torso the moment he touched down on the runway and bailed out of his jet.

He took a deep breath and turned slowly around to survey the damage. Until that moment, all he'd seen of his airplane was the limited view from the cockpit and narrow ladder coming down. What he saw steam and hiss on the runway made his knees go wobbly, his feet turn to clay.

"Oh, Gina," he touched the chest pocket of his flight suit where he kept his lucky silver dollar. "You and the boys almost won the lottery."

***

Feb. 27

Dear Tuck,

Thank God the war's over. President Bush made the announcement an hour ago. So strange to see all those Iraqi soldiers surrender in droves. So many old men and young boys, one guy wore Bermuda shorts and sandals. So much for Saddam's elite Republican guard and his Mother of All Battles.

I just got off the phone with Michelle. She said Sally went to bed with an ice pack after hearing the news. Apparently, this has spoiled her plans for "The Rapture." I guess the ground war didn't last long enough for Pastor to gather up his flock and head for the hills.

Your daddy called. Your mother went on the wagon an hour ago. The boys say "Hurry home." Me too. I've got cold beer in the fridge and my own brand of rapture waiting for you when you get here.

Be careful,
Gina

PART III

Too Many Ghosts

Chapter Seventeen

High noon, Good Friday, 29 March, Beauregard Air Force Base. They came in waves through the silver clouds, four-ship after four-ship after four-ship. In a ceremonious fly-by, the A-10s passed over the airfield where most of the base and half the city of Bolton lined the ramps that ran up and down the length of the taxiway. Far out on the horizon, a dark thundercloud kept its distance.

Tuck peered out the side of his canopy at an ocean of humanity below, with traffic bumper to bumper from the airfield to the front gate. Another line of cars snaked along on the road in and out of the back gate, near his quarters.

Tuck's radio squawked.

"Who do they expect, the Thunderbirds?" chuckled Killer in the plane in front of him. "You'd think it was Open House."

"Fuckin' A," Tuck shot back with an eye on his instruments. "Lap it up while it lasts, Captain. Americans are a fickle bunch."

Television trucks from all the major networks, with their

satellite dishes aimed toward space, scattered around the ramp. Someone painted a huge yellow ribbon on top of one of the hangers.

Tuck pitched out on downwind with a safe distance between Killer's plane in front and the guy behind him. The control tower warned them to expect severe cross winds.

After Killer landed, Tuck lined his A-10 up on the runway that grew bigger by the second.

"Home sweet home." He relished the moment his jet touched down on the pavement and rolled down the runway. He never thought he'd see the day when he would feel this happy to get back to Louisiana. He guided his jet down the center line and fought to keep it from going into the weeds from the cross winds.

Slowing, he turned his jet toward the taxiway, followed the slow-moving train of A-10s, and approached the crowd up and down the flight-line.

When Tuck got closer, he popped open the canopy and unclipped his oxygen mask. His tinted visor shielded his eyes from the sun. He taxied past the well-wishers who cheered and wiped their eyes behind the roped off area.

He gulped. Overcome with emotion, he watched the thousands of people from all walks of life who turned out to wave American flags and yellow streamers at their returning heroes.

Tuck didn't feel like a hero, but he did feel proud that the American people stood up and applauded their armed forces, instead of turning a cold shoulder like they did after Vietnam.

An entire high school marching band stopped playing and broke into a chant when his jet rolled by. "Hip hip hooray! Hip hip hooray."

Cymbals crashed and the drummers rat-tat-tatted. Twirlers threw their batons in the air and spun around to catch them. A fat kid on a tuba blew several loud notes like the trumpet blast of an elephant.

Tuck laughed and flashed them a hero's grin. Thrusting a gloved hand in the air, he saluted them with a victorious thumbs-up.

"It's good to be home," he cried. His voice lost in the wind and the savage screams of two dozen jet engines to proclaim their arrival.

Moments later, he pulled his jet into the chocks and waited for a crew chief to signal him to shut down his engines. After flipping a few switches, the ear-piercing wail died down and another crew chief scrambled under the belly and chocked the tires.

Tuck pulled off his helmet and caught his breath at the vision of a lady in red, breaking from the crowd. Her long, chestnut hair blew in the breeze.

Sweet Jesus.

Gina, decked out in a red micro-mini and matching pumps, flung her arms in the air like kite strings.

A feeling of warmth rushed over him. He beamed and watched her run. She abandoned all pretense of decorum or protocol and squealed like a lovesick teenybopper as she crossed the tarmac.

Michelle, two steps behind her in a pale-yellow miniskirt and sleeveless blouse, her shoulder-length hair flew like a golden veil. Her legs were bare and creamy from winter, and her boots, laced up in yellow, were all that remained of an era before the world turned topsy-turvy and sent them scattering in opposite directions: Michelle back to Galveston, Tuck to Saudi.

Clutching the boys' hands, Michelle scurried them along in new suits and bow ties. Their legs pumped to keep up. Tuck couldn't believe how much the kids changed in seven months.

He couldn't get out of his jet and down the ladder fast enough. With his helmet bag in hand, he bolted toward them, his G-suit girded his waist and thighs. A sudden gust of wind almost knocked Jesse down as the whole family crashed head-on into each other.

"Daddy Daddy," a chorus of voices exploded around Tuck. Dropping his helmet bag on the tarmac, he wrapped them all in his arms. They smelled like heaven and home and soap and shampoo and everything good in the world. He buried his face in their necks, their hair; he couldn't get enough of them.

After they all embraced, Jesse tugged at his dad's pant leg.

"We didn't get a doggy." He scrunched his face up at Tuck. "But Sissy came back."

Tuck scooped him up in his arms and smothered him with kisses. "I can see that." He reached to stroke Michelle's cheek.

He ruffled Austin's hair and stared at him a moment. "I see you've got your two front teeth."

Austin grinned and they all laughed.

Then Tuck swept Gina off her feet and twirled her around like a ballerina. She squealed with delight. He set her down and kissed her hard on the mouth.

A photographer from *Time* magazine approached. He had his press pass clipped to the pocket of his denim shirt. Tuck could tell by the grin on the man's face he already captured them on film.

"Welcome back, Colonel," the photographer said. He tipped his head at Gina and Michelle and grinned at the boys. "You've got a beautiful family. Can I get you folks to sign a photo release? In case we run your picture?" He held out a pen.

Tuck hesitated, uncomfortable.

"Oh, c'mon, ya old fuddy-duddy," Gina elbowed him in the ribs. "Do it for posterity."

Tuck frowned.

"Yeah, ya old fudgy-wudgy," giggled Jesse then clapped his dimpled hands in the air.

"Come on Dad," begged Michelle. "Don't you want to be famous?"

"Not really," Tuck grumbled.

Austin handed him the pen. “Dad, you’re outnumbered.”

Tuck signed the form and passed it back to the photographer.

“Thanks for your time, Sir,” the man said and handed him his business card. “If the editors use it, I’ll be in touch.”

“Fair enough,” Tuck smiled, dropped the card in the chest pocket of his flight suit.

The photographer turned and looked at Michelle. “Has anybody ever told you you look like a model?” His eyes zoomed in on Michelle like the lens of a camera.

“All the time,” she giggled then batted her lashes and flipped her hair back.

“Seriously. You should consider it.” The guy blinked at her one last time, then he moved off for more subjects.

Michelle stood with her mouth open. “I feel like Miss America.”

Tuck took Gina in his arms. “I missed you,” he murmured then slid his hand up her skirt and goosed her in front of the boys.

Gina laughed and wrapped her arms around his neck. “I missed you too.”

The wind picked up. Brushing hair out of Gina’s eyes, Tuck could feel the temperature drop and the sudden change of pressure in the atmosphere. He glanced at the boys, clinging to his legs for affection.

“Guys, we better head for cover.” A huge raindrop smacked his nose.

***

At an area designated for VIPs, Becky Spitz stood in the background and watched her husband pander to the media. By now she heard the ugly rumors: Bull was a coward, and his men knew it. Becky clutched her purse while her husband strutted back and forth in front of a host of television cameras and played the part of the big war hero. *Bull the actor*, Becky sniffed to herself. Feeling disgraced, she watched him puff out his chest like an ape. Now she knew how Lois Dennison must have felt, except Colonel Dennison had the decency to keep a lid on his shortcomings.

Becky took a deep breath and gazed skyward. She invested too many years in Bull's career to give up on him now. Besides, adultery was a tad easier to cover up than an act of cowardice. She squared her shoulders, gathered up her brood of three stocky boys and scurried them toward the squadron. The sky looked funny and she wanted to get inside before the wind trampled her new hat.

"You'll get to see your papa soon enough," she told them and ignored their backward glances and bustled them through the crowd. "Now, listen up, men. No one, and I mean no one is allowed to touch the food until the party begins."

"Yes, Ma'am," all three answered dutifully. They shuffled along behind her.

"Mama," her youngest tugged at her side, "did you really pay some lady a hundred dollars to bake that warthog cake?"

"She sure did," smarted Becky's thirteen-year-old son. "And it's made out of real hog guts too."

"Oh, it is not," Becky chided.

"Is too," blurted her middle son. "And it'll drip blood when she cuts it."

Becky steered them up the walkway in front of the 428th. Holding open the door, she shook her finger at the two oldest boys. "You two drop it, before I have a nervous breakdown."

Once inside, she rushed them into the rec room, which she spent hours decorating. "You boys go wash your hands," she shooed them away from the cake. "And for God's sake, don't touch anything. You know how your papa gets when it comes to his squadron."

***

By the time Wheaties landed, it started to sprinkle. Sylvia ran to meet him. *Hot dog*, but her little figure had blossomed.

A second later, his joy turned to horror when he watched Sylvia stumble, then double over, and crumple to the ground.

"Hold on, Darlin'," he yelped and broke into a sprint.

He scooped her up and whisked her off toward the sea of people corded off behind the yellow ropes.

Warm, wet fluid oozed down the front of his flight suit when he charged full speed ahead like a linebacker with the ball into the end zone.

Two Air Force medics leaned casually against the hood of an ambulance parked near the edge of the crowd, their backs to the ramp and eyes on the clouds rolling across the horizon.

"Yo, Driver, " Wheaties yelled.

The older medic whipped his head around.

"My wife's gonna have a baby," Wheaties said.

Both medics bolted upright and ran to get the stretcher.

Overhead, the first few rumbles of thunder churned the sky into a cauldron of black, smoky brew.

Strapped flat on her back, Sylvia groaned.

The medics loaded her in the ambulance and took off for the hospital. The wail of the siren obliterated the wind that whistled at the edge of the windows. The older medic drove while the younger one in the back prepped Sylvia for delivery.

"Turn off the siren." Sylvia tossed her head from side to side.

"Sorry," the medic said and pulled down her panties. "It's an evil necessity."

"The siren or this?" she snapped and pushed his hands away.

Wheaties pried her fingers loose from the medic's forearm. "Now, Darlin', this is no time to be modest. These guys do this all the time."

The young medic looked at Wheaties and broke into a sweat.

Halfway off the flight-line, they crept at a snail's pace through the crowd, before the whole sky opened up like the bomb bay of a B-52 and dumped its fury upon them.

The driver flipped on the windshield wipers and radioed the hospital. "Base One, Mobile Two en route..."

Sylvia moaned, the siren howled, and the rain beat down in driving sheets; hail, the size of golf balls, ping-ponged off the metal roof and littered the ground like shards of glass.

"Play me some music," Sylvia growled. "I hate that siren."

Wheaties leaned forward to poke his head into the cab. "Is this rig equipped with a tape deck?"

The medic hunched over the steering wheel, tried to see out the window.

"Yes, Sir," he said. "But to tell you the truth, Lieutenant, we usually don't listen to music when we have a patient on board."

"What, is it against the regs?"

"You could say that." The driver shifted in his seat

Sylvia got louder, more demanding in her discomfort. "I want music."

Wheaties inched closer.

"Look, Driver," he hung his head and tried to gain the medic's sympathy. "I know I'm a lowly lieutenant," his voice dropped to a conspiratorial whisper, "but I haven't been laid in seven months. I have a rash on my butt from the scratchy toilet paper they made us use over there, and my wife's back here fixin' to squeeze out a watermelon." He paused to let it sink in. "Now you think I give a hoot about some lousy regulations?"

"O-kay, Lieutenant," nodded the driver. "But I gotta warn you." He reached under his seat and pulled out a tape box. "All I've got is stuff from the seventies, Aerosmith, Boston, Kansas."

"Kan-sas," Wheaties roared and slapped the back of the vinyl seat in front of him. "Why, that's home, Driver. Crank'er up and let's boogie. Sylvia's gettin' restless."

***

Clouds rolled in, black and ominous. Prongs of lightning streaked across the darkened sky; thunder clapped and hail beat down with no warning.

"Follow me," Tuck yelled and pointed his head at the enormous aircraft hanger nearby.

Gina tore off her shoes and straddled Jesse on her hip. Michelle grabbed Austin and they scrambled into the thick of the crowd.

By the time they ducked into the hanger, drenched and soaking wet, a hundred other people followed.

They huddled together near the entrance to keep an eye on the storm. Tuck put his helmet bag down and took Jesse from Gina and bounced him on his hip. "You're getting too big for your mama to carry."

Trembling, Gina felt for Austin's head cuddled against Michelle. Instinctively, Austin wedged himself between the two women.

A siren pierced the air. Gina spotted the ambulance moving through the frantic crowd. The base hospital was half a mile away.

"I hope nobody got struck by lighting." She pulled Austin closer as a huge bolt of lighting zigzagged precariously close to them.

"Me, too," Tuck agreed. A loud crack of thunder overpowered his voice, followed by a loud ka-boom, like the noise of a jet breaking the sound barrier.

Gina flinched.

Jesse clamped his hands over his ears. "Go away," he lashed out at the sky. He buried his face in his daddy's shoulder.

"Hey, guys, lighten up. I made it home in one piece, didn't I?"

Austin nodded, his brown hair plastered to his forehead. "Yeah, Dad, and nobody spit on you like that homeless man in the newspaper."

Michelle stared out at the rain. "Same thing happened to Jesus. They spit on him, too."

Gina rubbed her arms to stay warm. "This storm wasn't in the forecast."

Michelle pushed wet hair out of her eyes. "It's like a big, black cloud snuck up out of nowhere and gobbled us up."

Gina shivered at the thought.

"Hey, Jesse. Know why it's raining?" Austin pulled away from her.

Jesse lifted his head off Tuck's shoulder. "'Cuz God's crying?" he blinked his green eyes at his brother.

"No, dummy," Austin laughed. "It's 'cuz he's giving Sarge a bath."

"Who's Sarge?" Tuck frowned.

"A dead dog, dummy," Gina hissed sideways and jabbed him in the ribs. "Remember?"

***

The ambulance got stuck in traffic. Half the population of Bolton tried to get off base and jammed Beauregard's narrow roads.

Kansas' *Point of No Return* played over the tape deck.

Sweat covered Sylvia's face.

"It's coming," she panted, and squeezed the sides of the stretcher. "It's coming out of my buuutt," she screamed.

Her face contorted to the point where Wheaties didn't

recognize his own wife. "Push, Darlin', push," he coaxed, pulled out a handkerchief and mopped her brow.

The medic moved in, his head disappeared between her legs.

Wheaties winced.

An earsplitting scream reverberated off the ambulance walls and drowned out the music, the rain, the hail...

"Push, Mrs. Wheaton," the young medic commanded her. "I can see the baby's head."

Sylvia grunted like a wild animal.

The song in the tape deck wound to an end. The rain and hail stopped, as if God snapped his fingers and said, "Enough."

A tiny head of coppery fuzz appeared then shoulders and swoosh, the baby arrived, and the medic held him up like a trophy.

Sylvia panted and tried to catch her breath while everybody stared at the baby.

"It's a boy," the medic declared, followed by the cry of a baby filling his lungs with air for the first time.

"Hot damn," Wheaties yelped. "I got me a son."

The ambulance lurched forward into the flow of traffic.

The next song filled the air with the haunting lyrics of *Dust in the Wind*.

Chapter Eighteen

Monday, April 15, Wynonna Sandford paused in front of the entrance to Kmart and admired herself in the window. She liked the way her new diamond necklace sparkled at her throat. Big Sandy said it made her eyes look like sapphires. Fluffing up her pile of curls, she felt like Marilyn Monroe. *Now all I need is a mink coat. That will come later, by golly, if I come up with another money-maker like Operation Purple Passion.*

Sales skyrocketed for the Fort Worth-based company before the last shipment of free goodies went over, right before Christmas.

Admiring her necklace one last time, Wynonna followed Big Sandy into the store. She thought of Ms. Katherine Tuttle's remark at the award banquet last week in Fort Worth. "Ladies, take it from a pro like Wynonna. It pays to be patriotic."

Inside the store, Wynonna smiled at the old lady in blue tennis shoes who handed her a circular. The woman's eyes lit up when she saw Wynonna's necklace.

While Big Sandy went to get them a shopping cart, Wynonna rifled through the circular, looking to see what was on sale. They killed time until they headed off to Rose Glen to make her afternoon rounds. On leave, Big Sandy promised to help her with deliveries if she went fishing with him when they finished.

They had to get Big Sandy a new pair of blue jeans because he'd worn a hole in the back pocket, where he kept his pouch of chewing tobacco.

They hadn't gotten very far when Big Sandy stopped dead in his tracks in front of her and blurted out, "Holy smokes, Schnookums! Looky there. On the cover of *Time*."

Wynonna stopped short, glanced up from an ad about foot massagers.

"Follow me," Big Sandy charged. He wheeled the nose of their shopping cart around and zoomed off to the nearest checkout.

Frowning, Wynonna stashed the circular in her purse and hoofed off after him. "Wait up," she called. The heels of her feet slapped against the top of her slides.

When she finally caught up with him, Big Sandy pulled a magazine out of the display rack and chuckled at something on the cover.

A pimple-faced checkout boy noticed them. "Check it out," he said, holding up another copy. "It's a couple from Beauregard. Says so right here on the cover."

"Good gravy," Wynonna crowed and snatched the magazine out of the kid's hand. "It's Colonel Crabby and Gina, carrying on like they just got married."

Against a backdrop of black clouds, Tuck and Gina Westerfield were caught in a passionate kiss while their three children huddled off to the side like guests at a wedding.

The checkout boy glanced at Wynonna and Big Sandy. "I take it you know these people?"

Big Sandy nodded enthusiastically. "They live next door to us. That's Colonel Westerfield. He's retiring today."

"Those fly-boys really kicked butt in Iraq!" exclaimed the

checkout boy. He gave Big Sandy the once over. "You a pilot too?"

Wynonna's eyes snapped up and glanced at her husband.

Big Sandy rocked back on his heels and cocked a half-crooked grin. "No, Sir," he hitched up his belt buckle. "I'm the guy that builds their runways."

"Can't fly without runways," the young man pointed out. He leaned across the counter and looked at the picture upside down. "Check out this chick in the yellow mini." He let out a loud wolf whistle.

Wynonna blinked, offended by his remark. "A chick," she heard herself say, "is a little yellow critter that runs around a barnyard." She squared her shoulders, gave him a sugary smile. "This here is a young lady."

The young man blushed and straightened back up. "I meant it as a compliment, Ma'am."

Big Sandy looked at the clerk and chuckled. "You'll have to excuse my wife." He draped an arm around Wynonna. "She gets protective when it comes to that little gal."

Wynonna's eyes misted. She bowed her head, stroked the cover of the magazine.

Big Sandy gave her a squeeze. "It's 'cuz of you she cleaned up her act."

Pulling a tissue out of her purse, Wynonna dabbed at the corners of her eyes, careful not to let her mascara run. She started to laugh. "I can't take all the credit, Daddy."

The clerk eagerly poked his head back over the counter. "What was she like before she cleaned up her act?"

Wynonna tilted her head in thought and allowed a sly grin. "Does the Bride of Dracula ring a bell?"

The kid thought about this for a moment, then started to laugh. "You mean she was a Goth? The black hair? The black lips and stuff? We had a couple of them in school. Pretty creepy if you ask me." He crossed his arms to study the picture. "Hard to believe. Considering she looks like an angel now."

Wynonna looked up at him, then back at the picture and thought, *My, how angelic Michelle looks in pale yellow, with her halo of blonde hair and skin the color of sweet cream.*

Wynonna felt something stir inside. For the briefest moment, she wondered, *Why did God take my baby or why couldn't Michelle have been my daughter?* Then she remembered all the mangy dogs and cats of this world that might have died without her help.

Big Sandy flipped through the magazine devoted to Operation Homecoming celebrations at several military installations around the country.

"Look here, Schnookums. It says, 'Louisiana Base Welcomes Home Desert Heroes.' Now how do you like that? Right here in a national magazine."

Wynonna pulled her eyes away from Michelle and turned to see what Big Sandy yakked about. The photo spread included a shot of an A-10 as it landed on the runway at Beauregard —Colonel Maple saluted a crew chief and stepped away from his jet onto the ramp, a black sergeant kissed a baby and an ugly kid with a mouthful of braces waved a flag.

Big Sandy turned a page and started to laugh. "It's the Big Kahuna himself, Colonel Spitz, running in full retreat. Pretty damn ironic if you ask me."

Wynonna stooped to get a closer look. The picture showed a big, white-headed pilot run toward the hangar, pass women and children in his quest to get out of the rain.

Wynonna read the caption out loud: "Unidentified A-10 pilot runs for cover in a freak storm that put a damper on Beauregard's festivities."

She and Big Sandy laughed heartily. By now everybody on base knew about the 428th's wartime coward.

"You reckon ol' Pedro worried about his hair?" Wynonna quipped.

The checkout boy cleared his throat. "Who's Pedro?"

While Big Sandy told him the story, Wynonna's eyes drifted to the other magazines on the display rack. Fingering her diamond necklace, she scrutinized the covers of *Glamour* and *Vogue*.

Michelle can do that, she thought, recalling the conversation they had last summer at her dining room table, when the girl said she wanted to model. Just last week in Fort Worth, Wynonna overheard Ms. Katherine Tuttle tell one of the national sales directors that Purple Passion wanted to update its image.

Clutching the magazine to her breasts, Wynonna hurried out of the store. With a face like Michelle's on the company's brochures, every woman in the country would be tempted to try Purple Passion's products.

Come next winter, Wynonna would snuggle up inside her new mink coat.

"Schnookums," Big Sandy hollered and chased her down in the parking lot. He glanced at the copy of *Time* in her hands, the one she'd forgotten to pay for. "They'll nab you for shoplifting. And what about my pants*?*"

***

At 1100 that morning, Lieutenant Colonel Bull Spitz stormed into the squadron and told his secretary to hold his calls. He came back from the BX, where he had to wrestle some enlisted swine for the last copy of *Time*. He slammed the door to his office and threw the magazine on his desk.

He dropped in his chair, feeling disgraced. He'd already seen the entire issue, front to back. Even though his name wasn't mentioned in the picture that showed him fleeing from the storm, people would know it was him. They would laugh and make comparisons between the photo and his performance in the war. Even his own men laughed at him now. He could feel it in their eyes, hear it in their voices whenever he walked through the squadron.

Mostly, he stayed in his office, awaiting orders. After the change of command next month, he would get the hell out of there. Becky had her sights set on D.C. He didn't care where they went, as long as he got a desk job.

General Dennison would take care of him.

The phone on his desk jangled. He snatched it up. "Irene, I thought I told you to hold my calls."

"Yes, Sir, but your wife is on line. She'd like to know what time you'll pick her up for Colonel Westerfield's retirement ceremony."

Westerfield? He completely forgot about him.

"Tell her to meet me at the hanger," he snapped.

He glared at the picture of Westerfield and his wife, caught up in an amorous embrace. Something you would expect from an airman or sergeant, not a senior officer who was supposed to show some dignity.

Spitz picked up the magazine, crumpled it in his beefy hands, and tossed it into the waste basket.

He would never understand a man like Westerfield, who could walk away from a promising career and thumb his nose at Dennison's offer to command the 428th, then look so goddamn, deliriously happy about it on his way out the door.

Spitz turned off the light and marched out of his office.

Irene looked up from her desk, where she polished an apple. "Have fun." She took a juicy bite as he walked past.

Before leaving the building, Colonel Spitz stopped by the men's room. No sooner had he unzipped his flight suit than he saw something that made his blood run cold. Taped above the urinal was his picture from *Time*. Someone drew a tail between his legs and wrote "PEDRO, headed south" in big bold print.

Word got around.

***

Tuck's retirement ceremony was over in fifteen minutes. Twenty years of his life wrapped up in the time it took him to get ready for work every morning.

He put on his sunglasses and marched from the big hanger, wearing his Class-A uniform and a chest full of ribbons. He carried a folded flag, a few more medals to hang on the wall, and the promise of a monthly pension.

In the parking lot, he loosened his tie, struggled out of his jacket, and regarded Gina with a grin. "Now all I need is a job."

"You'll get one." She slung her purse over her shoulder and took the flag from him.

Her voice is too chipper, too optimistic, Tuck thought.

So far, he sent applications to all the major airlines. So far, no one called. His pension would pay a few bills, but they couldn't live on it. He had no degrees beyond a B.A. from LSU in political science, specializing in Soviet foreign policy and with the collapse of the Soviet Union, his degree was as useless as a plane without wings.

He needed a flying job, and one that paid well. But flying below the radar to smuggle cocaine out of Columbia was out of the question.

Holding the triangular bundle in front of her, Gina carried the flag as reverently as a Girl Scout ahead of him.

Tuck took off his wheel hat and tossed it to Austin. "Here, kid, be somebody."

The round hat with a flat top flew threw the air like a Frisbee. Austin caught it by its bill and stuck it on his head. "Thanks, Dad."

"What about me?" Jesse frowned.

Tuck stooped and draped his jacket over the back of Jesse's shoulders. The jacket swallowed the boy. "Here, soldier."

Jesse brightened instantly and, together, the three traipsed off across the parking lot and hollered at Gina to wait up.

Once in the Bronco, Jesse chanted, "Daddy's retarded. Daddy's retarded."

"It's re-tired, you bozo," Austin eyed his brother with contempt. "Uncle Bo's the one that's retarded."

Tuck cringed and watched them in the rearview mirror. "He is not." He searched for the right words to describe his brother's condition. "He's not wired right, that's all."

"I believe the proper term is mentally challenged," Gina sniffed, folded her hands in her lap, and rested them on top of the flag. She sighed and looked out her window.

In the driveway at home, Austin threw off his seatbelt and lunged for the door. "Come on, Jesse. Tonight we'll sleep in the rent house."

"No more fire ants," Jesse jabbered.

Gina lifted him from his seat. Tuck didn't tell his son that fire ants weren't picky when it came to their choice of neighborhoods. Knowing Jesse, he'd figure it out soon enough.

Tuck went to unlock the house. "What time does the cleaning crew come? The inspection is Wednesday."

Gina bent over the flower bed to look for something. "Sometime after lunch." She reached into the marigolds and came up smiling. "Hey, Austin. Recognize this?"

She held out the crash helmet to Austin's Lego man.

The light on the answering machine flashed when they paraded into the kitchen. Except for the phone and an advanced copy of *Time* from the publishers, the countertops were empty.

The boys took off down the hall, their voices echoed from room to room as they tore through the house, opened and shut closets, and terrorized the dust bunnies.

"The house looks so lonely." Gina clutched the flag.

Tuck peeled off his sunglasses. "I thought you hated this place."

With the pictures gone, the furniture packed up, the throw rugs jellyrolled on the back of a truck, the house looked like any other empty set of government quarters in need of a good cleaning and a few small repairs. Nail holes gaped from every wall, in need of putty and a fresh coat of paint. Every nick and stain, every old cigarette burn from some previous occupant, now showed up on the kitchen counter.

The phone rang. Tuck snapped it up in hopes of a call about a job. Instead, another reporter wanted to set up an interview. Tuck politely declined. Since their picture came out in the magazine, everyone wanted a piece of them.

Even the boys grew tired of their instant celebrity. At the retirement ceremony, everybody wanted to pinch their cheeks and declare how photogenic they were.

As for Tuck, he found the whole thing embarrassing. He told Gina the photo looked like the cover from one of Jo-Ellen's romance novels.

Against the clickety-clack of the boys' dress shoes on the linoleum floor, Tuck picked up a pen and pressed the button on the answering machine.

"Hi, guys. Jeri Sweeney here. Long time no see. Fabulous photo. Tuck, glad you made it home safe. I finally heard from the Air Force. The SOBs have ruled Jeff's accident pilot error. Call me. Miss you guys."

Tuck swallowed hard, then bleeped to the next call.

"Dad, Gina, I'm home sick with the flu. Mom's over at Pastor's house. Guess he does income taxes too. By the way, Mom's really freaked about the magazine."

The next call was from Bo. "Tuuuc-ker, Mama says you a ce-leb-ri-ty now."

Smiling, Tuck went to the next call as the boys clattered past, headed out the door.

"Gina, it's Mom. You guys are the talk of the town."

The last call was from Gina's father. "Hello daughter and family. Your picture is the biggest thing to hit Llano since the cow mutilations back in the seventies. Got your taxes done I hope?"

No calls from the airlines. Hiding his disappointment, Tuck went to check on the boys.

Jo-Ellen Hawkins bustled up the driveway in her espadrilles and denim skirt, her pocketbook on her shoulder and a plateful of cupcakes in her hand. "I baked these myself," she bragged with a deep, sassy laugh. She passed Tuck a cupcake and waltzed into the house.

Wynonna scuttled over from next door. "Yoo-hoo," she called with Baby under one arm like a football. She followed Tuck into the kitchen where she waved around a copy of *Time*. "I'm fixing to make your daughter a star."

Tuck leaned back against the stove, licked the frosting off his cupcake. "And how do you propose to do that?"

With a gleam in her eye and a foxy grin on her freshly powdered face, Wynonna twittered, "Well, Colonel, I reckon I'm gonna have'ta make that ex-wife of yours my new best friend."

"Good luck," Gina snorted, taking the plate of cupcakes from Jo-Ellen. "The woman's a beast."

Wynonna looked down at her furry dog, nestled safely in her arms. "Mama knows how to tame nasty ol' beasts, doesn't she, Baby?" The dog whimpered and stared at the cupcakes.

Tuck sidled up next to Wynonna and said, half teasing, "Are you referring to your mutt there? Or Big Sandy?"

The doorbell dinged just as Wynonna looked up at him and grinned. "Actually, Colonel, I was referring to you."

Tuck chuckled and went to answer the door.

Krystal breezed in with a bottle of champagne and a stack of paper cups. "Can't stay long," she announced, shoved the bottle toward Tuck and hurried into the kitchen. "We got a big trial coming up," she added and lined the paper cups in single file on the counter. "Too much to do, so little time." She smiled. "Not that I'm complaining, of course."

Tuck went to open the champagne. Austin and Jesse scrambled into the kitchen and spotted the Pekingese. "Baby," they cried and ran to greet the dog. "Did you come to tell us goodbye?"

With a twinge of guilt, Tuck watched Wynonna kneel and let the boys pet the dog.

"She's so cute," Jesse giggled.

Austin pointed toward the dog's genitals. "Baby's not a girl, you knucklehead."

Even Tuck had to grin. While he listened to them chatter, the image of another dog floated up from the back of his mind. A shaggy little dog named Scooter. Tuck tried to suppress the memory, to rebury the image under the silt of time where all childhood things must go in the end. But Scooter kept coming, tail wagging, running up to greet him after school.

Tuck closed his eyes, remembered the day Scooter went to the vet and never came home.

"You okay?" Gina snuck up behind him.

Tuck's eyes flew open.

Gazing at the boys, he popped the cork and poured the champagne.

The doorbell rang. The cleaning crew arrived.

Later, as Tuck went to unplug the phone and haul it out to the car, American Airlines called to set up an interview.

***

"I hope no one calls the cops," Sylvia said.

While she nursed the baby, Wheaties rolled out of bed in his T-shirt and gym shorts, pulled on his flight boots, and boogied into town.

In the seat beside him, cold and silent inside its padded

case, lay his trumpet.

After passing the Grimes' place, where Tony would get up in another hour to fly, Wheaties took a left at the next intersection, turned off his headlights, and killed the engine. Using the light from a nearby street lamp to guide him, he coasted the rest of the way in, easing the Jeep to the curb in front of a white-columed house.

Grabbing his trumpet and keys, he scurried across the yard to the side of the house. After scaling the fence, he slid the trumpet up on the roof, pulled himself up, then tip-toed across the shingles.

With trumpet in hand, he jumped up and down to make as much racket as possible. Once the porch light flicked on the yard below, Wheaties puckered his lips, drew the horn to his mouth, and blasted away with his own special rendition of *Reveille*—a jazzed-up version of the famous bugle call that roused soldiers from their bunks.

When he finished, he slid off the roof and came around to the front of the house. "Hey, Driver," he yowled when he saw Tuck in his underwear, half asleep. "Did I wake you?"

Every dog in the neighborhood barked.

Tuck stood in the middle of the front porch, holding the door open. "Get your ass inside. You're gonna wake the dead."

In the kitchen, piled with boxes, Tuck popped open a beer and handed it to Wheaties. The clock on the microwave said 3 a.m.

"Sir, you got anything to eat?" Wheaties held his stomach.

Tuck tossed him a box of Animal Crackers.

"How's it feel to be an old retired fart?" Wheaties crunched down on a buffalo.

Tuck yawned, stretched his arms. "Lieutenant, ask me that in the morning when you get ready for work and I'm sleeping in."

Laughing, Wheaties ate another cookie and gulped his beer.

Gina stumbled into the kitchen, rubbed her eyes. "Wheaties, do you have any idea what time it is?" She squinted at him with one eye.

Wheaties glanced at his watch and whistled. "Sorry about the hour, Ma'am." Grinning, he reached over and slapped Tuck on the back. "I couldn't let this self-respecting fighter pilot of yours leave the Air Force without a good, old-fashioned roof-stomp."

Gina gave him a sleepy grin and turned to go. "Don't eat all the tigers," she yawned, then shuffled back to bed. "They're Jesse's favorite."

After Gina left the room, Wheaties set the box of cookies down and peered at Tuck. The punchy grin, the good-natured expression were gone.

Tuck knew what was coming.

"Who did you tick off before the war?"

Tuck drained the last of his beer and stared at the lieutenant in his gray gym shorts and flight boots. Wheaties cracked his knuckles.

Tuck told Wheaties everything. He didn't give a shit anymore.

Wheaties doubled over with laughter, holding his gut. "I knew it, Driver," he gasped and snorted for air. "I knew you had somebody's dirty pictures." He wiped his eyes with his T-shirt. "I just didn't know whose."

Tuck squeezed his beer can and tossed it in the trash. "You should have seen the look on Linda Garrett's face when her head popped up," Tuck said matter-of-factly. "She looked like a gopher caught by surprise."

Wheaties threw his head back and laughed hysterically.

He held up a half-empty beer. Clearing his throat, he recited what was known as The Fighter Pilot's Toast. "Here's to me in my sober mood, when I ramble, sit and think. And here's to me in my drunken mood, when I gamble, sin and drink..."

Tuck popped open another beer and joined in.

"But when my flying days are over and from this world I pass..."—their deep, sudsy voices filled the room, and together they belted out the rest of the toast— "I hope they bury me upside down, so the world can kiss my ass."

From the far end of the house, Gina's voice bellowed up the hall, "Will you two yahoos knock it off out there. Some of us are trying to sleep."

Chapter Nineteen

Tuesday, June 25, Wheaties and Killer took off from the base at 0900. The sun was out for the first time in days. Except for a few puffy clouds scattered across the horizon, the sky was clear.

"Hot damn, Driver," Wheaties hooted over his radio as they headed west. "It's a beautiful day for flying..."

Killer, in the lead plane, chuckled back: "Roger that, big guy. Let's rock and roll."

***

As Jo-Ellen Hawkins walked into the base emergency room with her bloody finger wrapped in a dishcloth, she didn't yet know that the painful memories of another time would come back to haunt her before the sun went down.

With a foolish grin on her face, Jo-Ellen approached a female med-tech. "I cut myself slicing a watermelon," Jo-Ellen said as the med-tech led her to a small examining room to check her vitals. "I'm forty-four years old. Boy, do I feel stupid."

"It can happen to the best of us." The technician took Jo-Ellen's blood pressure. She ushered Jo-Ellen out to the waiting room, where Jo-Ellen joined a tired-looking young woman wrestling with a fussy toddler. "The doctor should be with you shortly."

"Looks like you've got your hands full," Jo-Ellen looked sympathetically at the young mother. "Wished I could give you a hand, but I accidentally cut myself this morning."

***

At a small fishing shack on the banks of the Sabine River, on the Louisiana/Texas border, a young black boy emerged from the doorway with a can of corn and a fishing pole. After walking down to the river, he stuck a few kernels of corn on his hook, plopped it in the water, and waited for his bobber to sink.

He adjusted his baseball cap when two A-10s screamed in low over the treetops in front of him.

He looked up and like always, got that warm, fuzzy feeling that made him tingle and feel good all over like the time he caught a baby catfish and let it go.

He liked to watch the warplanes, especially now that he saw them on TV.

The first A-10 passed over the river just as his bobber went under. He started to look away, to pull in his fish. But then the second plane flew over, lower than the first. Something went wrong. Terribly wrong. The airplane wobbled then pitched sharply to the right. Clutching his mouth in horror, the boy watched the airplane crash wing first into a thicket of trees down the bank from him. A huge fireball ignited the sky and rolled out across the water.

The boy dropped his pole and ran screaming to the cabin where his father cooked breakfast.

***

Sylvia Wheaton put the baby down for a late morning nap

and finished putting the groceries away. After taking a package of pork chops out of the freezer, she settled back on the couch with a library book and a bag of raw carrots.

She crunched into a carrot, turned a page, and sunk further into the couch. Nap times were the only time she had to herself anymore, not that she minded. She loved that little baby sleeping in the next room more than life itself.

***

Jo-Ellen's ears perked up when she heard a commotion coming from the nurse's station. Med-techs scrambled out of nowhere and rushed to leave.

"What's going on?" Jo-Ellen asked nervously as a technician scurried past.

"Can't say, Ma'am." He barely gave her a nod.

Jo-Ellen glanced across at the young mother who seemed unconcerned by this sudden buzz of activity.

"Probably just another exercise," the woman yawned, bouncing her cranky toddler on her knee.

"So soon after the war?" Jo-Ellen got up to check. She marched to the nurse's station. "Excuse me. Is there an exercise going on?"

The technician hesitated, averting her eyes. "If you'd like to take a seat, the doctor will be with you shortly."

Jo-Ellen swallowed. "A plane hasn't gone down, has it?" she blurted, then broke out in a cold sweat.

The technician ignored her.

Jo-Ellen gathered herself up and fled for the door, her injured finger forgotten in her rush to escape.

Using her good hand to drive, she white-knuckled the steering wheel all the way home. She needed to talk to Buzz, the one man, the only man, who could help her forget the terrible scene unfolding in her mind:

She was back at the high school in Fort Walton Beach

on her first teaching job, and she read Shirley Jackson's *The Lottery* to her fifth period class. The principal stuck his head in the door and summoned her into the hallway. When she saw Mike's commander with his hands folded, she fled.

She ran down the hall, bumped into one of her own students, Skippy, who was out on a hall pass. She barely noticed the freckle-faced boy as she bolted for the nearest exit. In the courtyard, under the droopy leaves of a giant weeping willow, she collapsed.

Then Mike's commander, along with the principal, stood there and patted her gently on the back. Together, both men told her the news she didn't want to hear, her handsome new husband, a second lieutenant right out of pilot school at Vance, crashed his A-1 Skyraider into the Gulf of Mexico. He hadn't survived. As she curled into a ball to disappear, Skippy and the rest of her fifth period class came outside and dropped down next her and refused to leave her side, even after the bell rang for them to go to their next class.

***

A lone A-10 circled high above the smoldering wreckage like a mother eagle looking for her lost eaglet. The flight lead couldn't leave his downed wing man.

Later, after a search-and-rescue helicopter zoomed in over the blackened banks of the Sabine River to confirm no sign of a chute, the lone A-10 pulled out and headed back to base.

By early afternoon, the Air Force had the entire crash site roped off. Beauregard's new Chief of Safety scoured the area, took notes, joined by dozens of others sent out on detail to tag pieces of wreckage and the pilot's remains.

Nearby, Air Force Photographer Brian Crenshaw, on his first accident detail, mumbled a silent prayer as his camera clicked away. Airman Crenshaw felt sick to his stomach. Why hadn't anyone warned him that jet pilots don't come through high-speed crashes looking like anything human.

His camera zoomed in, captured the image of a smoldering, charred hulk, all that remained of the pilot's decapitated torso. A chunk of something lay nearby, possibly a boot. He took pictures of everything, from every angle, leaving out nothing, then he went into the bushes and vomited.

Later, when he learned the identity of the dead pilot, he wept openly, and longed for the sweet accompaniment of a trumpet that would never come again.

***

The white-top full of blue suits pulled in front of the little house on base. Four grave-looking officers stepped quietly from the car and began the somber procession up the front walk.

The notification team included Colonel Maple, the wing commander, Chaplain Lawrence with his Bible, the flight surgeon gripped his black bag of sedatives, and Captain Tony Grimes clenched his fists at his sides.

Engrossed in her book, Sylvia jumped when the doorbell rang. Getting up from the couch, she laid the book down and went to answer the door almost tripping over the diaper bag.

When she cracked open the door, she blinked in confusion. Why did these four solemn faces stare back at her with that god-awful look in their eyes?

It took her a second to comprehend before it sunk in: The death angels. There to deliver such devastating news that it took four of them to do it; a doctor, a chaplain, a friend, and the boss.

She gripped the doorknob, felt the blood drain from her head.

God in heaven, she thought and braced herself for the blow.

She looked straight into Tony's eyes, so dark and sad

she didn't need to ask questions. Still, her words tumbled out, unbidden and stiff. "Don't candy coat it, Tony," she heard herself say, "What happened?"

Tony glanced at Colonel Maple, who nodded for him to go on. Tony stepped forward and took Sylvia's hand, his dark, boyish face hard as iron. "His plane went down over the Sabine River," he said gently, in that soft-spoken way of his. "He didn't punch out, Syl..."

Sylvia felt her throat close up. She closed her eyes, tried to picture him falling out of the sky. Flying into the ground. Silent forever.

At that moment, if the baby hadn't started to cry, a loud reassuring song that life would go on—she could've sworn that all the music in the world died along with her husband.

Feeling numb, she took a deep breath and invited them in. "Have a seat while I get the baby," she heard herself say. "He's like his old man. All he thinks about is food."

She started to walk away then turned. "Wheaties is a good pilot," she blurted defensively in a voice she didn't recognize. "He's been around airplanes all his life. Don't try to tell me it's pilot error." She shook and glared at them in defiance. "It was mechanical. Wheaties is too good."

Colonel Maple folded his hands and looked across the room at her. "Mrs. Wheaton," his voice full of compassion. "It's too soon to tell anything at this point. That's for the accident board to decide. Nobody is jumping to conclusions, as far as I know."

She glanced at Tony, who nodded his head and said softly, "Colonel Maple's right, Sylvia. It's too soon to tell." Then he lowered his eyes. Chaplain Lawrence had his Bible propped open on one knee, ready to read her some bit of comfort.

But nothing could comfort her now except Wheaties' strong arms wrapped tightly around her, telling her everything would be all right.

The doctor started to get up, and Sylvia put up her hand.

"Save your tranquilizers for somebody else," she snapped, then turned on her heel to get the baby, bolstered by an electrifying anger that took over her body.

***

Screwing the cap on the jar of mayonnaise, Gina realized how much she missed hearing the twelve o'clock whistle, the roar of jet engines by the flight-line, and the feeling of community she always took for granted when they lived on base.

"Face it," she told herself and put away the mayonnaise and mustard, "You miss it."

Sucking on a lollipop, she assembled the turkey sandwiches and wrapped them in Saran Wrap for their picnic with Jo-Ellen at the base playground in twenty minutes.

On the way there, Gina wanted to cruise by their old set of quarters on Bayou Way and check out the new family. Jo-Ellen said she had seen three of the cutest towheaded girls running through the sprinkler the other day. After the picnic, Gina wanted to stop and check on Sylvia and the baby. She hadn't seen them for a while.

As Gina packed the last sandwich into the cooler, she heard a knock at the side door by the carport.

"Somebody's at the door, Mommy," said Jesse, coming into the kitchen. He tried to stuff too many Hotwheels into the pockets of his shorts.

"You and your brother go potty and get in the truck." Gina closed the cooler. "I'll go see who it is."

Jesse looked up at her and giggled. "You sound funny Mommy and your tongue's all purple."

Hefting the cooler, Gina bustled past him to answer the door. "Make it snappy," she barked over her shoulder. "We still have to stop for ice."

In the utility room, she set the cooler down and opened the door. Krystal, her cheeks stained with tears, stood on the other side of the screen door and rubbed her palms against a white linen skirt.

Gina spit out her lollipop, holding it at her side. "Krystal, what in the world's wrong?" She unlatched the screen door and held it open.

Krystal stood frozen next to the Bronco. "Tony just called me at work," her voice sounded so shaky Gina could hardly understand her. "Oh God, Gina. It's Wheaties."

Gina frowned. "Wheaties?" For a fraction of a second she didn't understand. Yet Krystal's face said it all. Suddenly, Gina doubled over, gasped for breath. "No! No! No!" she stammered like a stubborn child.

Before Krystal could catch her, Gina stumbled back against the doorjamb and slid to the floor. The screen door banged against her leg with a thud but she didn't care. "Not Wheaties," she wailed bitterly and shook her fist up at God. "They just had the baby."

Jesse stood over her now with his face puckered up in fear. "Mommy don't cry," he whimpered. Frightened, he looked up at Krystal. "What's wrong with Mommy?"

Krystal bent over, cupped his face in her hand. "Mommy's sad, Honey. Real sad."

Gina wanted to reach out to her son, to draw him near. But all she could see through a blur of tears, was Wheaties in her new kitchen in the middle of the night, gobbling Jesse's Animal Crackers.

***

In Fort Worth, Texas, Tuck slowly put the phone back on the jack. He sat on the edge of the bed and stared at the wall. He could hear the sound of water running through the pipes of the motel. Someone took a shower or flushed a toilet.

Behind him, spread over the bed, lay the training manuals he'd been studying when Gina called.

"Sylvia wants you to give the eulogy," Gina said after breaking the news. "Can you come home? Even for a day?"

Tuck told her he would see what he could do.

Outside his door, he heard a couple of new hires like himself talking when they walked past his room. They were either on their way to the flight academy or coming back to the motel to study. That's all American's new hires did in their rooms: sleep and study. And fantasize about their wives and girlfriends. A whole year of pilot school, crammed into two months.

He got up to comb his hair. He had class in an hour.

He stood in front of the medicine cabinet and let the water run. With a wounded sigh, he allowed himself to feel nothing but self-contempt.

"You weak-dick son-of-a-bitch," he glared at himself in the mirror. "Why were you the lucky bastard chosen to beat the odds?"

He waited for an answer.

But the guy in the mirror didn't respond.

***

At 2 a.m., Tuck crawled into bed next to Gina. He'd been on the road six hours, after a full day of academics.

"You feel good." He slid next to her in the dark.

Curled up in a fetal position facing the wall, Gina twisted around and rested her head on his chest. "You know what Austin told me last night when I tucked him in?"

"What?" Tuck yawned, exhausted.

Gina rose up on one elbow. "He told me two clouds bumped into each other playing tag, and Wheaties got caught in the middle."

Gina laid her head back down and shivered.

Tuck pulled the sheet around her, kissed the top of her head. "How's Jesse? You think he's too young to understand?"

"He understands plenty," Gina mumbled. Her fingers played with the downy gray hairs on Tuck's chest. "He says God needed Wheaties in heaven to take care of Sarge."

Tuck couldn't help but chuckle.

"Of course Austin had to butt in and tell Jesse how stupid he was. He said: 'Jesse, you knucklehead, why would God need Uncle Wheaties when he's already got Jesus and Colonel Sweeney taking care of Sarge? Uncle Wheaties was supposed to help Miss Sylvia change diapers.'"

Tuck pressed Gina closer. He didn't know whether to laugh or cry.

Gina let out a sigh. "The boys are only four and seven, and they're already thinking about death."

Tuck yawned to avoid Gina's last remark. "How's Sylvia?"

"Okay, I guess. Jo-Ellen stayed with her last night."

"Jo-Ellen's probably the best person," Tuck said, "considering she's been through it herself."

Gina's voice quivered when she spoke. "Jo-Ellen says Sylvia catches herself looking out the window for Wheaties, half-expecting to see him come driving up with a big grin on his face and say, 'Hey, Driver, what's all the fuss about?' Jo-Ellen told her it's normal; part of the grieving process. She said it takes awhile for the finality to sink in."

"It's only been a couple of days," Tuck said.

Gina looked up at him in the dark. "Reminds me of that story about a faithful dog who doesn't know its master is dead. Poor old dog sits by the door day after day and waits for him to come home. But the master never does."

"Gina," Tuck held his breath. "There's something I've never told you."

Gina tensed up.

"I never expected to live this long," he admitted and caressed her hair.

He told her what the silver-haired colonel told them that

first day of pilot school back at Willie: one in three would die at the controls of an airplane.

Gina didn't say anything for a minute. Finally she whispered, "Sylvia's all alone."

Tuck realized it was easier for Gina to think about Wheaties' widow than the fact Tuck bared his soul to her, revealing his deepest fear, something she tried to drag out of him for years. Now she couldn't handle it.

Tuck cradled her head in his hands and stared at the ceiling. "No she's not. She has the baby."

Even in the dark, Tuck could see the blades of the ceiling fan swish round and round, cooling the air over their bed. When he was a kid, he pretended ceiling fans were the propellers to P-51s, Japanese Zeros, and German Messerschmitts.

After Gina dozed off, Tuck rolled over and closed his eyes. Like so many times before, all he wanted to do was drift off into oblivion and forget any of this happened. Forget about Wheaties. Forget about Sweenedog. Forget about everything that happened over the past year. But tonight, the ghosts were restless, and they danced around in his head.

Chapter Twenty

"...old men will dream dreams,
...young men will see visions."
Joel 2:28

Tuck shivered, feeling out of place behind a big, wooden pulpit. Pulpits were for preachers, not pilots.

Rubbing his hands to stay warm, he confronted an empty sanctuary, dappled in sunlight. Beams of colored light streamed through the stained glass window to his right, yet the inside of the base chapel felt as cold as a meat locker.

He glanced at his Rolex, 1300 hours. He was supposed to deliver the eulogy, but there wasn't another soul in sight. It didn't make any sense, neither did the airline uniform he had on, complete with white shirt and tie. As far as he knew, he was still going through training.

Puzzled, he took a deep breath and stared into space. Any minute now the chapel doors would swing open and mourners would flood in.

Once everyone was seated, the boys from the squadron

would march up the aisle, single file in their dress blues. With eyes of steel, they would fall in behind the first row of pews and stand shoulder to shoulder—four rows of rock-solid, unflinching flesh—and wait until Sylvia had been escorted in and seated in the pew directly in front of them.

Only then would they sit down, all at once and in perfect formation.

After going over the whole thing in his mind, Tuck looked down at his notes.

—Devoted husband; doting father; superb pilot

Somewhere in the distance Tuck heard a door squeak.

He expected to see Chaplain Lawrence or the others, but there was no one.

He glanced at his notes again, unable to concentrate. In the frigid air of the chapel, he wriggled his fingers and toes to get the circulation going.

An eerie silence descended over the room. A silence so alive and powerful, the hackles on the back of his neck stood up. He broke out in a cold sweat. A second ago he froze; now he burned up. Blood flowed back into his fingers and toes, and he gripped the lectern, aware of another presence in the room, a presence that materialized out of nowhere, daring him to look up.

He lifted his head, and halfway expected to see God Almighty or the Devil himself.

Scattered throughout the sanctuary were the men he mothballed so far back in his head he forgot about them. Until now; Chuck, Lurch, Rick, Lou, Roscoe, Harry, Burger, too many to name. John, Ed, Floyd, Ralph, Dave... Tuck felt overwhelmed.

Still, he marveled at the sight of them, at their cocky grins and unburned skin, at their limbs attached to their torsos, their torsos attached to their heads as if they stepped into the squadron snack room for a munchie or waited to go fly.

Their physical appearance baffled him the most.

Not one looked dead. Yet Tuck knew they were. He went to their funerals. He went to their crash sites. He picked up their pieces and dropped them into plastic bags, and then he went on with his life.

He leaned forward to get a closer look. The room swayed. The tilting motion rocked him from side to side across the length of the platform, before he slammed into the side of a wall, and fell back dazed. His belly flopped to his knees, sling-shotted up through his gut and bottomed out somewhere near his throat.

In a stupor, he pulled himself up and fought his way back to the pulpit. In one powerful, heaving shudder, everything lurched to a halt.

Tuck heard a crackling noise from the stained glass window. He watched in quiet fascination as the white dove in the glass fluttered to life and flew off and left a gaping hole in the glass.

A second later the rest of the window crashed down.

Stunned, Tuck groped his way across the room. The air from outside gushed in, hot and dank. The carpet, littered with shards of colored glass, crunched under his feet. Below him, the beautiful stained glass window lay shattered beyond recognition.

"Shit happens."

Tuck jerked his head around at the sound of the voice.

There, lurking in the shadows stood Larry, the lackadaisical lieutenant they nicknamed James Dean back at Willie. Larry, with his curled-up sneer and slicked-back hair, flew his T-38 into the ground a week before graduation and took his instructor with him.

Larry slouched against a wall smoking a cigarette. "I'd ask you to give me a ride back to the cemetery," he blew out a lungful of smoke, "but there wasn't enough of me to bury."

Tuck trembled, unnerved by his remark.

Looking half-cocked, Larry flicked his ashes on the floor then faded away.

"What's with the airline uniform?" somebody cranked up at the back of the chapel.

Casting his eyes in that direction, Tuck spotted the dark-haired major they called Bronco kicked back in one of the pews, nursing a long neck. He had on cowboy boots and jeans, his feet propped up on the pew in front of him. Exactly the way Tuck remembered him at the going-away party in Alaska, before Bronco flew off to Norway and barreled into the side of a mountain.

Tuck stared at him and said, "Fifteen lousy feet, Bronco, and you would have cleared the rocks."

"With all due respect, Sir, you look like a glorified bus driver," Bronco snickered then floated away.

Tuck grabbed the lapels of his new uniform and longed for the comfort of his baggy old flight suit.

"You're looking mighty fine there, my good man, for a civilian," said a voice as smooth and rich as the finest liqueur.

Tuck swallowed. He knew that voice anywhere. A voice as charming and elegant as the man himself. A voice snuffed out in a smoking hole over eastern New Mexico. Sam, the smooth-talking ladies' man they called Shaft back at Cannon AFB.

Tuck's eyes snapped forward to the first pew in front. Sam, the proud son of a former Tuskegee airman, stood tall and dapper in a tailored flight suit and the chunky gold ID bracelet he never took off his right wrist. Tuck stared at the bracelet, and recalled the day Sam's F-111 shot into the ground like a giant dart, and changed Tuck's perception of death forever.

Tuck had been a young captain back then, sent out on detail to search through the rubble. While Sam was pretty much splattered over a cattle rancher's land, mixed in with the cow patties and prairie grass, the bracelet survived the crash, without even a scratch. Tuck found it dangling on a Yucca plant, no more than a hundred yards away from the hole. When he plucked it up, he squeezed it, trying to milk it for the elusive answer to the question on everybody's mind that day: What the fuck happened out there on final?

"Sam?" Tuck grappled for the words. "You're dead."

Sam tipped his head back and gazed up at Tuck. His sable-colored eyes crinkled in amusement. "No shit."

He turned his back on Tuck and melted away.

Tuck dropped his head in his hands, and grieved all over again.

When he looked up, Wheaties floated through the air, his body intact. Cradled in his muscular arms was a baby with a crownful of coppery hair.

After landing feet first, Wheaties kissed the baby, placed him on the altar, and walked over and stood in front of the first row of pews. Facing the crowd, he yanked off his helmet, held it against his chest and yelped, "Hey, Drivers. Guess I won't be needing this anymore." He took a step forward, released the helmet with one powerful toss and sent it rolling down the center aisle like a bowling ball.

"Strike," somebody cried from the back of the chapel and the whole place went up in cheers.

Wheaties took a bow, winked at Tuck, then poof, he was gone. The baby wriggled alone on the altar, its tiny fist batting the air. Tuck gaped at the infant's little pink mouth opened in surprise.

While Tuck contemplated what to do with the baby, a familiar voice yodeled over his shoulder. "It's show time, boys."

Tuck spun around, startled at the image before him.

Jeff Sweeney hung upside-down on the cross, like a bat hanging from a cave. A wedge of blond hair shot out the top of his head, almost touching the floor. His jolly blue eyes twinkled with amusement. He looked at Tuck from the foot of the cross and grinned.

"Hey, movie star. How did you manage to get your sorry ass immortalized on the cover of *Time?*"

Immortalized? Tuck rolled his eyes. What an absurd

thing to say. "It's just a magazine. Come on, Sweenedog, you know I hate the limelight."

"Hate it?" Sweenedog snorted. "Shit, Bubba, I'd swap places with you in a heartbeat if I had one."

Tuck closed his eyes, pinched the bridge of his nose.

He felt guilty. Guilty for being alive. "I'd trade everything I have if I could bring you back to life." He lifted his head, and gazed awkwardly at his friend.

Sweenedog paused then let out a throaty laugh. He banged the back of his head against the foot of the cross as if it was the funniest thing he'd ever heard in his life. "Who do you think you are, Tuck, Jesus Christ?"

Tuck hung his head, speechless.

When he looked up, the cross was empty, and the wistful sound of a woman's voice filled the air. Tuck turned, drawn to the voice. Deep inside of himself, in a part he no longer acknowledged, he realized the voice belonged to a woman he once loved.

He found his ex-wife in the choir loft, strumming her guitar and wearing an orange caftan with beads around her neck.

Sad-eyed and gloomy, Sally gazed out at the sanctuary, and sang an old antiwar folk song about dead soldiers, flowers, and cemeteries.

She stopped playing, glared at Tuck, and pointed a disapproving finger at him. "When will you ever learn, Tucker? The military breeds nothing but death."

Tuck shuddered then felt himself float away. He felt like water as he flew through the air, his body fluid and interchangeable.

The next thing he knew he groveled around in a bed of wet grass clippings.

He'd been there before, he realized, feeling a trickle of fear. He was a child again, a boy of seven or so, and yet a part of himself remained grown. He seemed part-boy, part-man, and it felt only natural that he could move back and forth between the two.

Back in the cemetery in Rose Glen, he suffocated in the stench and decay all around him. A field of sunflowers grew nearby, their droopy heads toppled over, half dead in the heat. Even then, they were taller than he.

His eyes locked on the fat slab of marble, the long name that would take him years to grow into. A name that defined him, told him who he was. His name was as much a part of him as his own soul.

Yet here it was, written among the dead.

Even as his eyes latched onto the shorter name below it, on the name *Fossy*, his mind struggled to understand. Fossy was a dead man. A grandfather ripped apart under the blade of some horrible saw, then tossed into the earth to rot.

Every shaft of fuzz on the nape of his scrawny neck stood up. Death was all around him, in him, no escaping it. He felt the rush of warm pee soak his crotch, and he sprung from that place, and screamed for the nearest exit.

Halfway across the cemetery, a hand came out of nowhere and grabbed him from behind. Too scared to scream, he turned, and stared helplessly into the empty eye sockets of his grandfather.

"Tucker Foster Westerfield," the sawmill monster gurgled his name. "I've come for you, boy."

The last thing Tuck saw was a statue of Jesus, standing there all stone-cold and silent, helpless to save him.

***

Friday afternoon, 28 June, 1400 hrs, mourners spilled out of the base chapel in groups and pairs and left behind the comfort of refrigerated air and the serenity of indoor lighting. The second the eye-stinging glare of afternoon sunlight hit their red, swollen eyes, many slipped on sunglasses or raised their programs to their brows, and squinted painfully beneath the words printed there: *In Remembrance...*

By the time Tuck and Gina emerged, a large crowd gathered on the lawn. They waited for the flyby, the formation known as The Missing Man.

The eulogy went well, Tuck thought as he stuffed his notes in Gina's purse. Once, when Tuck said something funny about Wheaties, Wheaties' Uncle Matthew—f resh from the cockpit of his Piper Pawnee, parked at a small airstrip north of the base—blurted out in a crackled voice, "That's my boy, Roy. Always the life of the party."

Seeing that most of the shady spots were already staked out, Tuck steered Gina toward the nearest unclaimed tree, a dainty mimosa growing off by itself. Huddled under the flouncy canopy that was more for show than shade, Gina fanned herself with her program while Tuck scanned the crowd for Sylvia.

Earlier, before the service, Sylvia cornered Tuck in the choir room. "These were Wheaties' first set of wings. I want you to have them."

Tuck suggested she keep them for the baby, but Sylvia shook her head and pressed the wings into Tuck's hand. "Why fill little Matt's head with the same dreams that killed his father? Let Uncle Matthew carry on the legacy. Right now, I see it only as a curse." She apologized and walked off, leaving Tuck by a closet full of limp choir robes, staring at a cold set of wings.

At last he found her, in a simple black street dress and low-heel shoes, under the shade of a large oak tree, greeting visitors. She appeared thinner, but strong. As strong as the live oaks and pecan trees dotting the church yard.

Sylvia seemed to console everyone else instead of the other way around. During the eulogy, when Tuck told the crowd how Lieutenant Wheaton saved his life in the Gulf, Sylvia bowed her head and smiled. At the same time, Wheaties' new squadron commander, a round man with an easy grin and a bald, shiny pate, who'd only known Wheaties a month, wept.

Like a gallant old soldier, Uncle Matthew stood next to her, tall and gangly and weather-beaten, in a ratty flight suit stripped of its insignia. A patch on his chest pocket said *Wheaton's Flying Service.*

With a flick of his wrist, Uncle Matthew ran a bony hand through his thinning crop of strawberry-blond hair and shook a few hands now and then. Mostly he gazed at the sky, like a farmer checking for rain, although the only thing he would find up there at the moment was an ocean of air.

Tuck wondered what Uncle Matthew would think of Sylvia's decision to hand over Wheaties' wings? Years ago the old man had to bury his brother; today he had to bury his brother's son.

A mosquito buzzed in front of Tuck's face. He swatted it so hard in midair that it was dead before it hit the ground. He noticed Bull and Becky Spitz come out of the chapel. After the Spitzes offered their condolences to Sylvia, they turned and split for the parking lot, not even staying for the flyby.

Tuck wasn't surprised.

He fixed his eyes on the sky. Although the roof of the chapel blocked his view, he knew that somewhere off in the distance, out on the horizon, a four-ship of A-10s would join up.

"This all feels like a bad dream," Gina said in a voice as dull and heavy as the stagnant air. She swished her program back and forth in front of her face.

Tuck didn't say anything.

Sweating and uncomfortable in a navy-blue suit he wore to his job interview, the frayed edges of last night's dream lingered in his mind. The memories of the people he dreamed about would haunt him the rest of his life.

Their job was to help him remember.

He stared at the crowd.

Tony and Krystal Grimes stood nearby, along with Buzz and Jo-Ellen Hawkins and a couple of second lieutenants who

Tuck didn't recognize. A cluster of pilots from the 428th huddled around a short, stocky captain. Killer, Wheaties' wingman, appeared visibly upset.

With trumpet in hand, a white-gloved airman stepped into the open and played *Taps*. At the first brassy, mournful note, Sylvia Wheaton stiffened, her head held high. When the music was over, she gazed skyward into the blue, her eyes unprotected by sunglasses, her face the color of bone.

Seconds later, a low rumble sounded in the distance, then grew louder, like a storm building to the south. Tuck threw his head back—eager, hungry—and waited for the ear-splitting vibrations that would break him in two.

Suddenly, the heavens exploded in a rapturous thundering chorus, as a four-ship of A-10s screamed overhead at five hundred feet.

As the four jets passed wing tip to wing tip over the top of the chapel, number three pulled up and away, leaving a blue hole in the formation to symbolize a fallen comrade. Higher and higher, the lone A-10 climbed, silhouetted against the blue Louisiana sky, like an iron cross flying into the sun.

Over the roar of the jets, Uncle Matthew cried out next to Tuck, "Go get 'em, kiddo." He turned and winked at Tuck, the corners of his dry, leathery lips crinkled back in a wry grin.

Tuck looked up at the sky again, and struggled to stave off the hot, salty tears that welled up in his eyes. "I thought I could save him," he heard himself say. "I thought I could teach him how to stay alive."

A tear dribbled down Tuck's right cheek. Then another.

Uncle Matthew reached over and gripped him by the shoulder. With a crazy, faraway laugh in his voice, he said, "The boy knew the risk. Let him go."

Epilogue

June 2011, Fort Worth, Texas. The sun felt good on Tuck's back as he sat in front of the bay window, working on his latest project. He pushed up his reading glasses and concentrated on the model airplane in his hand. He painted the crew member's name on the left side of the F-111, beneath the plane's canopy. He'd been retired from American Airlines for three years, and built models to keep his mind active and his fingers nimble.

At sixty-three, he had more hair on his legs than he did on his head. Except for the slight love handles that lapped over the belt of his walking shorts, he was in good shape, mostly from doing yard work.

For once, he had the whole kitchen table to himself. Gina was in downtown Fort Worth, having lunch with Wynonna Sandford, the CEO of Purple Passion since the turn of the century. Big Sandy had been dead for years, cut down by a tiny spot on his tongue that turned out to be cancer.

Austin was at MIT, working on a postgraduate degree in aeronautical engineering. Jesse owned a Harley-Davidson shop in Port Arthur, Texas, and rode with Bikers for Christ on weekends. When Jesse's eyes went bad in the late nineties, his dream of flying did too.

Michelle ran a boot camp for troubled teens in Galveston, Texas, with the fortune she made from modeling. She never married. Her mother, on the other hand, ran off with Pastor after the war and hadn't been heard from since.

Tuck put the finishing touches on the name Lt. Col. Jeff "Sweenedog" Sweeney when the doorbell rang.

He glanced under the table at Wolf, an old German shepherd that followed Tuck everywhere. Wolf's ears spiked up.

"Well, boy, let's go see who it is."

Tuck put down the paintbrush and went to answer the door. Several steps behind Tuck, Wolf's hips swayed stiffly from arthritis.

When Tuck opened the heavy, beveled-glass door, he faltered a bit as he took a step back in the marbled entryway.

A broad-shouldered young man with red hair and freckles grinned nervously from the long front porch that Tuck and Gina's Realtor had called the verandah.

Tuck opened his mouth, but nothing came out.

The young man hesitated then stuck out his hand. "Colonel Westerfield? Cadet Matthew Wheaton, Sir. A & M class of 2013. Just passing through on my way to ROTC summer camp at Sheppard Air Force Base."

Rattled by the young man's appearance, Tuck took the cadet's hand and sputtered awkwardly, "Forgive an old man his frailties, but for a minute there, son, I thought you were your father."

Wheaties' son glanced at the wooden planks of the porch then grinned sheepishly. "My mama said you have something for me?"

In a flash, Sylvia's voice came back to Tuck. "Why fill little Matt's head with the same dreams that killed his father?"

Tuck bowed his head. So Sylvia learned the same lesson Tuck's own mother learned: If you try to deny a child his dreams, he'll work twice as hard to achieve them.

Fumbling for words, Tuck ushered the kid inside and excused himself for a moment. In the master bedroom, Tuck retrieved a small, antique box from his walk-in closet, opened it up, and gaped at the contents.

The musty smells of age came up to greet him. The box, made out of walnut and lined with felt, had been passed down to Tuck from his father. The box contained items from several generations: a bullet from the Civil War, an old pocket knife, now rusted, a Rotary Club pin, an assortment of cufflinks, his father's wedding band.

Pushing those aside, Tuck picked up two other items and stared at them a long time before he put the box up.

When he came back into the living room, Cadet Wheaton and Wolf were getting acquainted.

"He's ferocious, Sir." The dog wagged his tail and enjoyed the rubdown. "Did you raise him from a pup?"

"Nope. He showed up on our doorstep one day and we couldn't get rid of him." Tuck made a face. "My wife swears he's the reincarnation of a dog named Sarge. But that's another story."

Cadet Wheaton laughed and looked around. "Is your wife here? My mama told me to tell her hi."

"No. She's hobnobbing with the CEO of Purple Passion. Your mother would remember Wynonna Sandford. Her dog use to crap in my yard back at Beauregard."

Wheaties' son chuckled and stood up. "Beauregard. That was my dad's last base."

Tuck stepped forward and opened his palm. "Your mother asked me to hang on to these after your father died. Guess she wanted to protect you is all."

Upon seeing his father's wings, Cadet Wheaton let out an ancient sigh older than his twenty years. He turned the wings over in his hand and clenched his fist around them.

His voice was raspy when he finally spoke. “Sir, what’s the secret to staying alive in this business?”

Tuck looked at Wheaties’ son, his throat dry. “Secret? There is no secret, son. Just blind luck.”

Tuck took out his lucky silver dollar—the one dated 1935—and placed it in Matthew Wheaton’s hand. “This was in my grandfather’s pocket the day he died. Only thing on him that wasn’t mangled in the accident.”

Tuck told him what had happened. “Poor bastard must have suffered. At least in an airplane, it’s quick.”

The kid stared at the coin. “In God We Trust,” he said, looking up slowly. “Did you fly with this?”

Tuck shoved his hands into his pockets. “For over forty years. My grandmother gave it to me on my sixteenth birthday. The day I started flying lessons. Good luck, kid.”

From the mantel, the old clock that ticked away for a hundred years chimed three o’clock.

“Thank you, Sir. I’ll cherish it always, along with these wings. I better hit the road.”

With Wolf at his side, Tuck watched Wheaties’ son get into an old Jeep and drive away.

In the back of his mind, Tuck thought he heard his grandmother chuckle, “Oh, Tucker Boy. Granny was wrong about ghosts. They’re everywhere.”

“You are a mist that appears for a little
while and then vanishes… *~James 4:14*

ABOUT THE AUTHOR

Native New Mexican Kathleen Rodgers followed her husband from base to base during his twenty-year career as an Air Force fighter pilot. The mother of two grown sons, she lives in Colleyville, Texas, with her husband and Chocolate Lab, Bubba.

Her work has appeared in Family Circle Magazine, Air Force, Army & Navy Times, and Because I Fly, a poetry anthology by McGraw-Hill.